## CORRECTIONS

On FIGURE 1C (p. xviii) and 1D (p. xix) for "genus" read "family"; for "genera" read "families."

# THE VASCULAR PLANTS OF INDIANA:
# A COMPUTER BASED CHECKLIST

## Theodore J. Crovello

*Department of Biology*
*University of Notre Dame*
*Notre Dame, IN 46556*

## Clifton A. Keller

*Department of Education*
*Andrews University*
*Berrien Springs, MI 49104*

*and*

## John T. Kartesz

*Biota North America*
*2202 Ridge Road*
*McKeesport, PA 15135*

*The American Midland Naturalist*
*Notre Dame, Indiana 46556*

*and*

*University of Notre Dame Press*
*Notre Dame & London*

Library of Congress Cataloging in Publication Data

Crovello, Theodore J.
　The vascular plants of Indiana.

　Bibliography: p.
　Includes index.
　1. Botany — Indiana.　I. Keller, Clifton A.　II. Kartesz,
John T.　III. Title.
QK159.C76　1983　　　　581　　　　　　83-10024
ISBN 0-268-01923-1 (University of Notre Dame Press)

# Contents

# Introduction

Indiana is a key floristic and biogeographic area for several reasons. The northern two thirds of the State was once covered with glaciers, while the southern third was not. Today it serves as the boundary among several major North American biomes, including grassland, woodland, and forest. Lindsey (1966) edited a detailed summary of the natural features of Indiana to celebrate the sesquicentennial of the Indiana Academy of Science, and Lindsey et al. (1969) summarized the remnants of natural areas in the State.

Over forty years have now passed since Deam (1940) published his detailed flora of Indiana as it was known at the time. In the interim systematic botanists have studied in greater detail many of the genera and species found in Indiana. The result of such efforts has been a great increase in our understanding of each species and of their interrelationships. While such progress is important and valuable both to basic and applied science, researchers and students of the flora of a region like Indiana are left without an adequate information source. An up-to-date inventory or checklist of all known plant taxa is a prerequisite for any other systematic, ecological or biogeographic study involving Indiana, be it basic or applied.

## PURPOSE

Our goal has been to create a checklist of all native or naturalized vascular plants known to grow (or to have grown) in Indiana. We have made our checklist as up to date and as accurate as possible. This means that: the list should include all taxa reported to grow in the State; its names should reflect the latest nomenclatural changes; it should allow easy reference to Deam's (1940) Flora of Indiana, the standard reference work for the State; it should contain as few spelling errors as possible; and authors of taxa should be uniformly and correctly cited. Finally, the checklist should be created and maintained in a computerized system that permits easier error correction, modification, and production of subsequent revisions than would be possible by purely manual means.

## FLORAS ARE DYNAMIC

The flora of an area like Indiana is constantly changing, as is our assessment of it. The latter always lags behind the former, and thus is incomplete. Actual floristic change is due to additions and extinctions of taxa to the geographic area in question (a taxon is a general term referring to any formally defined set of organisms, be it at the subspecies, species, genus, or other levels of the taxonomic hierarchy). But our assessment is incomplete not only due to such actual floristic change, but also to both our incomplete past knowledge and inadequate current monitoring of the State's flora.

What is the actual number of species of vascular plants in Indiana, and which ones are they? This is difficult to answer because it involves a consideration of immigration, extinction, synonomy, level of collection, and the reliability and completeness of the last generally published flora of the State. The problem can be more clearly stated in a prose equation (Crovello and Keller, 1981).

The number of species actually present today in Indiana equals
the number in the last general flora (Deam, 1940)
*PLUS* the number of nomenclatural births since Deam (1940)

*PLUS* the number of new species collections since Deam (1940)

*PLUS* the number of immigrating and introduced species established since Deam (1940)

*PLUS* the number of newly evolved species (for completeness!) since Deam (1940)

*MINUS* the number of nomenclatural deaths since Deam (1940)

*MINUS* the number of actual State extinctions or extirpations since Deam (1940).

From the above equation, it should be obvious that we rarely will ever know for sure the number of species present in a given geographic area. We are reminded of the valuable concept of taxonomy as an unending synthesis (Constance, 1964). This concept applies also to biogeography. Recognized in this light, the problem of the number of species in an area becomes an exciting challenge rather than a consuming frustration.

## INFORMATION FOR THE CHECKLIST

The major sources of information for the present checklist are: Deam's (1940) Flora of Indiana; a series of articles published since 1940 in the Proceedings of the Indiana Academy of Science which reported new State and county records; two editions (Swink 1974, Swink and Wilhelm 1979) of the Flora of the Chicago region, including the seven counties in northwestern Indiana; and additional new State records received from members of the Indiana Department of Natural Resources and from others. For the last source just mentioned, a correctly identified voucher specimen had to be deposited in a readily accessible herbarium. We used Kartesz and Kartesz (1980) as our source of correct names.

The immediate stimulus to produce a computer-based checklist was the result of much of the data becoming available in computerized form as a byproduct of the detailed floristic classification study by Keller (1979). In addition, this checklist coincides with the longterm goals of the Indiana Biological Survey of the Indiana Academy of Science (Crovello, 1979; Crovello and Keller, 1981).

## PREPARATION OF THE CHECKLIST

The following paragraphs describe how the checklist was created. Such detail serves at least two important functions: to better inform users as to how a name was included or excluded; and to provide sufficient details to allow others to carry out a similar process in the production of computer-based floristic or faunistic checklists for other geographic areas. In the following paragraphs "keystroke" means to keystroke data into computer-compatible form, or into the computer directly.

1. Keystroke names of those taxa in Deam (1940) that also had county distribution data. For each such name the Deam map number (and occasionally a letter) was also captured and became that taxon's nameber. A nameber is a combination of digits and letters used to identify a set of information, an object, character, or computer user (Levin, 1970:13).

2. Keystroke taxa in Deam (1940) that had no maps. Each such taxon was assigned the next available nameber, usually that of the preceding taxon in Deam with the addition of the next available letter.

3. Keystroke the 66 taxa which were previously not known to occur in Indiana and which were reported in a series of articles since 1940. These appeared in the Proceedings of The Indiana Academy of Science. Information was provided to us by John Bacone (Indiana Department of Natural Resources).

4. Keystroke any taxa in Swink (1974) that were not found in Deam (1940). While some of these names were new State

records, others were only nomenclatural changes of taxa already reported in previous works. Any taxon recorded from Swink (1974) was assigned a nameber. The following guidelines were used in the assignment of a taxon nameber for each taxon recorded from Swink (1974):

   a. If the genus is already represented in Deam (1940), assign the new species the nameber of the species in Deam alphabetically preceding it, and the next available letter.
   b. If the genus is not used by Deam, find its closest genus according to Gleason (1963). If not in Gleason, use Fernald (1950) or Dalle Torre and Harms (Anonymous, 1958). Assign the new taxon the nameber of the last species in Deam belonging to this neighboring genus, and the next available letter.

5. A draft of the checklist which integrated names from all of the above stages was compared with the manual card files of John Kartesz's checklist of vascular plants of North America. While his card file was the source of Kartesz and Kartesz (1980), the cards contained additional information on synonymy. Kartesz made several types of changes, including: correction of nomenclatural "grammar", e.g., elimination of subspecies "typica"; spelling and author corrections; addition of authors for the Swink (1974) names; and changes in taxonomic category levels. Even more importantly, he indicated real changes of taxa, including: substitution of the correct binomial for one erroneously used by Deam; updating nomenclature from Swink (1974); splitting of one taxon into two or more; and integration of two or more taxa into one.

6. Special computer programs were written to automatically check for certain types of errors. For example, were all map numbers captured from Deam (1940)? As an unexpected by-product, our program designed to check that all map numbers were represented revealed that Deam did not use map numbers 1520, 1887, and 2151. Currently, the only reasonable explanations appear to be that he just forgot to assign these, or perhaps he had

assigned them to a species which he subsequently eliminated from the final draft of his manuscript.

7. A computer card was punched for every taxon whose entry had an error in it, be it a misspelling, an incorrect name, etc.

8. The computer cards from step 7 were merged with a punched card deck of all the vascular plant taxa of Indiana. At the same time, the erroneous cards they replaced were removed. Although more sophisticated means of data set correction were available in our computer system, we chose to use cards and to carry out manual replacement because of the large number of records that had to be replaced (about 1,000). We also took this opportunity to proofread each record once more.

9. Using a series of sophisticated computer programs created by Keller, the data file of approximately 2,400 taxon names was changed from upper case only to both upper and lower case according to proper taxonomic usage.

10. Other computer programs verified that all taxa had authors and that all references to the same author were spelled the same.

11. Using the computer, all duplicate names were removed from the checklist. This program also permitted another opportunity to verify proper spelling of names. Information on duplicate namebers for the same taxon was preserved for later use. While not important for production of the Indiana checklist, it is essential information for revision of geographic distributions among Indiana's 92 counties.

12. All binomials in Swink and Wilhelm (1979) were keystroked into the computer, and automatically compared with our checklist. Any of their binomials that were not in our checklist file were printed out and evaluated one name at a time.

13. Genus names from Kartesz and Kartesz (1980) were keystroked. These served as data for still another program to verify the correct spelling of genus names.

14. At several later stages of this project we received reports of new Indiana State records from several sources. These were

assigned a unique taxon nameber which also allows us to quickly determine the source of the new record. In general, claims for new Indiana taxa were only accepted if a voucher specimen were deposited in an easily accessible, recognized museum, *and* if the specimen's identification was verified by a qualified person. These new records were integrated into our master files.

15. Special computer programs were used to create the several lists contained in the body of the checklist.

## RELIABILITY OF THE CHECKLIST

At least three different concepts of reliability are germane to taxonomic checklists. First, are the taxa nomenclaturally up to date and correct? For this we rely completely on Kartesz and Kartesz (1980). It is important to note that they did not compile their checklist of all North American vascular plants by themselves. Rather they enlisted the aid of over 250 taxonomic specialists in different groups of vascular plants. The authors then diligently integrated the opinions of these specialists with the most recent published work in the various taxa. We do not believe a sounder basis for a more correct and up-to-date nomenclature is possible today.

A second type of reliability concerns the following question. If a name appears in our checklist, is the taxon *really* found in Indiana? It may not really occur in Indiana because records of its occurrence in Indiana could be based on misidentified specimens. For example, species X is reported to occur in Indiana. But it is based on one or several specimens that someone misidentified. They really belong to species Y. Such mistakes have been minimized by basing occurrences only on either the professional publications mentioned or by requiring deposition of a voucher specimen and its professional identification for any new State record.

The third type of reliability involves another question. If a name does *not* appear on our list is it *really* in Indiana? Given the large geographic size of Indiana, the relatively little current collecting, and the continuing movement of species (especially weed species), the possibility will always exist that some taxa actually living in Indiana have not yet been found. While relatively few taxa native to Indiana are still being discovered for the first time, it is not clear whether this is because few species remain to be discovered, or because few systematists remain to discover them!

## CONTENT AND ARRANGEMENT OF THE CHECKLIST

Since our goal is to make this checklist of vascular plants as useful and as valuable as possible, it is first divided into three parts: the Pteridophyta (ferns and fern allies); the Gymnospermae (conifers); and the Angiospermae (flowering plants). Within each of these the taxa are arranged alphabetically by family and then alphabetically by genus and species within each family. This is the same arrangement as in Kartesz and Kartesz (1980). Our list integrates all taxa recognized in Deam (1940) that are still recognized today, as well as nomenclatural changes and records of new species found in Indiana since then. The following information is given for each taxon (reading from left to right on any page in the checklist):

1. Page in Kartesz and Kartesz (1980) on which the taxon's *family* begins.

2. Page in Kartesz and Kartesz (1980) on which the taxon's genus begins.

3. Taxon Nameber. The majority are simply the map number associated with each taxon that Deam (1940) mapped. A nameber followed only by a letter has several meanings. If it is for a taxon that also has a 1 or 2 in the previous column (indicating native or naturalized), then it most likely is a taxon that is found in Deam (1940) but for which he supplied no map. If no 1 or 2 is found in the previous column, then the taxon is from Swink (1974).

xii

In some cases the nameber consists of a number and letter followed by another digit. In this case the last digit indicates that a more recent source first reported that the taxon is found in Indiana. Current code number suffixes are as follows:

"1" Taxon first reported in a series of articles since 1940 which appeared in the Proceedings of the Indiana Academy of Science. We received a list of new taxa from John Bacone (Indiana Division of Nature Preserves) in correspondence dated March 28, 1978.

"2" Taxon first reported in Swink (1974).

"3" Taxon first reported by professional staff of the Indiana Department of Natural Resources, especially its Division of Nature Preserves and its Natural Heritage Program.

"4" Taxon first reported in Swink and Wilhelm (1979).

"5" Taxon first reported from a source other than those just described.

4. A one digit code indicating whether the taxon is considered native (1) or adventive (2). If a 1 or 2 is absent, it helps to indicate the source of the taxon as a new state record.

5. The complete taxon name, including its author(s). Our nomenclature agrees with Kartesz and Kartesz (1980).

## SUPPLEMENTARY LISTS

In addition to the major checklist, three other lists are included:

1. A list of taxa which appeared in Deam (1940) and *also* have: undergone nomenclatural changes; been reduced to synonymy; or for which Deam (1940) had given an incorrect author. For each taxon in this list the nameber and name from Deam (1940) is given in upper case. This is followed by the currently accepted nameber and name in proper upper and lower case format. This list is especially valuable for readers who might be

interested in a name that Deam (1940) used, but which is no longer accepted. They need simply note the taxon's nameber used by Deam (1940). Our list of changes is arranged according to Deam nameber. Readers are reminded that the large size of this list does not indicate major flaws in Deam's work as much as it shows the amount of systematic research that has been carried out since 1940.

2. A list of families known to occur in Indiana. For each family the following information is provided: its beginning page number in our checklist; the family name; and its number of species and genera in Indiana and North America.

3. An alphabetical list of families and genera reported to occur in Indiana. Each taxon is preceded by the page number of our checklist on which it begins, and succeeded by its page number in Kartesz and Kartesz (1980). For easy reference this list of families and genera is found at the back of this publication.


## STATISTICAL SUMMARY OF INDIANA'S VASCULAR FLORA

The Indiana checklist includes 2,265 species of vascular plants. They are distributed in 761 genera of 150 families for an average of 2.9 species per genus. According to Kartesz and Kartesz (1980) in North America north of Mexico the same 150 families contain 20,521 species in 2,649 genera, an average of 7.8 species per genus. Thus for these 150 families, Indiana's vascular flora includes 29% of the genera and 11% of the species found in North America north of Mexico. Indiana contains only 0.5% of the land area of North America north of Mexico and 1.2% of the land area of the coterminous United States. These data coupled with floristic data like the above can serve as a heuristic context to formulate hypotheses about what controls the average area occupied per genus or species in North America.

Of the 150 vascular plant families reported from Indiana, 24 are represented in North America by only one genus, while 66 are represented in Indiana by only one genus. Among the 150 families one is represented in North America by only one species (Lardizabalaceae: *Akebia quinata*), while in Indiana 36 families are represented by only one species. Figure 1 summarizes the frequency of numbers of species and of genera per family, both in Indiana and in North America.

The following ten families have the largest number of genera in Indiana (number of genera in parentheses): Poaceae (75); Asteraceae (72); Brassicaceae (36); Fabaceae (32); Lamiaceae (30); Apiaceae (28); Liliaceae (26); Scrophulariaceae (24); Rosaceae (21); and Caryophyllaceae (18).

The following ten families have the largest number of species in Indiana (number of species in parentheses): Asteraceae (255); Poaceae (223); Cyperaceae (222); Rosaceae (101); Fabaceae (93); Brassicaceae (72); Lamiaceae (70); Scrophulariaceae (59); Liliaceae (49); and Ranunculaceae (45).

We also examined the correlation between all pairs of the following four characteristics of the 150 families: number of genera and species in North America and in Indiana. In addition the coefficient of determination (the correlation coefficient squared) indicates the amount of variance that the two characteristics have in common. It can range from zero to one. Conversely, the remaining variation is not shared, an indication of how much is due to unshared causes. The correlation coefficients (with coefficient of determination given as a percentage in parentheses) are:

North American genera and Indiana genera: .94 (88%)
North American genera and North American species: .94 (88%)
North American genera and Indiana species: .83 (69%)
Indiana genera and North American species: .90 (81%)
Indiana genera and Indiana species: .87 (76%)
North American species and Indiana species: .91 (83%)

With a sample size of 150, all the above correlations are very significantly different from zero. Of special interest is the decrease

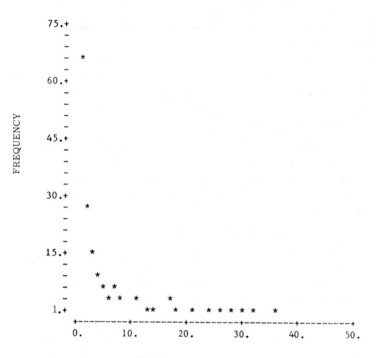

NUMBER OF GENERA PER FAMILY IN INDIANA

FIGURE 1A. Truncated frequency distribution of the number of genera per family in Indiana. Two families have over fifty genera per family (their number ranges from 72 to 75).

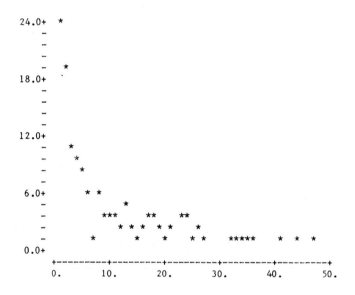

NUMBER OF GENERA PER FAMILY IN NORTH AMERICA

FIGURE 1B.   Truncated frequency distribution of the number of
genera per family in North America.   Eleven families have over
50 genera per family (their number ranges from 60 to 346).
Only the 150  families reported from Indiana are plotted.

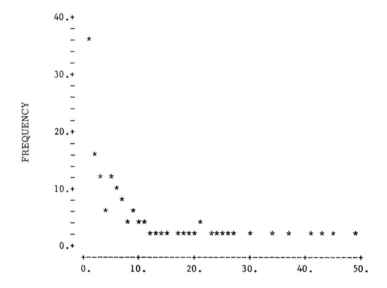

NUMBER OF SPECIES PER GENUS IN INDIANA

FIGURE 1C.  Truncated frequency distribution of the number of
species per genus in Indiana.  Eight genera have over fifty
species per genus (their number ranges from 59 to 255).

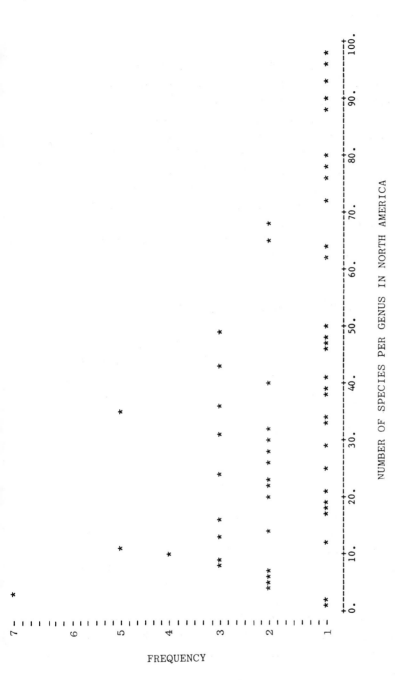

FIGURE 1D. Truncated frequency distribution of the number of species per genus in North America. Thirty eight genera have over 100 species per genus (their number ranges from 107 to 2,687). Only the 150 families reported from Indiana are plotted.

of 12% in the coefficient of determination as one goes from North American genera and species (88%) to Indiana genera and species (76%). Possible explanations include the following: 1) Indiana totals are based on smaller totals and thus the estimates of its parameters may vary more (but 761 genera and 2,265 species are not small numbers!) 2) Indiana still may be reflecting a "floristic founder effect", due to two thirds of it having been glaciated only about 10,000 years ago; or 3) a combination of Indiana's relatively small total size and the relatively few types of environment available within the State. More study is needed to determine the correct explanation, perhaps through the use of new methods of quantitative biogeography (Crovello, 1981).

## INDIANA'S RARE PLANT SPECIES

Bacone and Hedge (1980) published a list of endangered and threatened vascular plants of Indiana. They grouped such plants into eight categories after consideration of factors like the number of known sites, vulnerability and type of its habitat, its status in surrounding states, etc. A revision of the list was published by Bacone, Hedge, and Crovello (1981).

## FUTURE CHECKLISTS

Since our knowledge of what species are found in Indiana will never be complete, readers are asked to contribute to the next edition of the Indiana checklist. Should you believe you have found a taxon previously unreported for Indiana, please send a voucher specimen to one of the nearby herbaria which is a major repository for the State's flora. These include those at the following institutions: Department of Biology, University of Notre Dame, Notre Dame, IN 46556; Department of Biology, Indiana University, Bloomington, IN. 47401; and The Morton Arboretum, Lisle, IL. 60532.

Please collect a voucher specimen *only if* the future of the plant population will not be threatened by your activity. Record relevant information, including: date of collection; county name; exact location to nearest tenth of mile, or more if possible; altitude; names of other species at the location; size of population; your name; and your specimen collection number for this specimen (if you maintain collection numbers). Be sure to collect some of each relevant part of a plant (stem, leaves, flowers, fruits, rhizomes).

## ACKNOWLEDGMENTS

We acknowledge the University of Notre Dame and Andrews University for the use of their computing and other facilities. Without such help the project would not have been possible. We thank Doctor Charles Heiser for allowing access to collection records in the Herbarium at Indiana University, Bloomington, and all people over the years who have actively been involved in searching for and verifying new Indiana State records. We especially thank John Bacone, Director of the Division of Nature Preserves of The Indiana Department of Natural Resources. His cooperation, enthusiasm, and sincere interest in the plants and environments of Indiana is an inspiration to us all.

# LITERATURE CITED

Anonymous. 1958. Register zu De Dalla Torre Et Harms Genera Siphono-
gamarum ad systema Englerianum conscripta. Akademische Druck —
U. Verlagsanstalt. Graz, Austria. 568 pages.

Bacone, J. A. and C. L. Hedge. 1980. A list of endangered and threatened
vascular plants in Indiana. Proceed. Indiana Acad. Sci. 89:359–371.

Bacone, J. A., C. L. Hedge, and T. J. Crovello. 1981. The status of Indiana's
rarest plants; a revision of the list of endangered and threatened
vascular plants. Proceed. Indiana Acad. Sci. 90:385–387.

Constance, L. 1964. Systematic botany — an unending synthesis. Taxon
13:257–273.

Crovello, T. J. 1979. The Indiana Biological Survey: An unending synthesis.
Proc. Indiana Acad. Sci. 88:332–341.

Crovello, T. J. 1981. Quantitative biogeography: an overview. Taxon
30(3):563–575.

Crovello, T. J. and C. A. Keller. 1981. The Indiana Biological Survey and rare
plant data. In: Morse, L. E. and M. S. Henifin (eds). Rare Plant Conservation: Geographical Data Organization: 133–148. The New York
Botanical Garden. New York.

Deam, C. C. 1940. Flora of Indiana. Indiana Department of Conservation.
Indianapolis. 1,236 pages.

Fernald, M. L. 1950. Gray's Manual of Botany. Eighth edition. American
Book Company. New York. 1,632 pages.

Gleason, H. A. 1963. The New Britton and Brown Illustrated Flora. Hafner
Publ. Co. New York. In three volumes.

Kartesz, J. T. and R. Kartesz. 1980. A Synonymized Checklist of the Vascular Flora of the United States, Canada, and Greenland. The University of North Carolina Press. Chapel Hill. 498 pages.

Keller, C. 1979. Quantitative Techniques for the Determination of Phyto-
geographic Regions. PhD Dissertation. The University of Notre Dame.
Notre Dame. 338 pages.

Levin, I. 1970. This Perfect Day. Random House. New York. 317 pages.

Lindsey, A. A. (ed.). 1966. Natural Features of Indiana. Indiana Academy
of Science. Indianapolis. 597 pages.

Lindsey, A. A., D. V. Schmelz, and S. A. Nichols. 1969. Natural Areas in
Indiana and Their Preservation. Indiana Natural Areas Survey.
Lafayette. 594 pages.

Swink, F. A. 1974. Plants of the Chicago Region. Morton Arboretum. Lisle, Illinois. 474 pages.

Swink, F. A. and G. Wilhelm. 1979. Plants of the Chicago Region. Revised and expanded with keys. 922 pages.

# The Vascular Plants of Indiana

ADIANTACEAE

| | | | | |
|---|---|---|---|---|
| 1 | 1 | 0038 | 1 | Adiantum pedatum L. |
| 1 | 1 | 0037 | 1 | Cheilanthes lanosa (Michx.) D.C. Eat. |
| 1 | 2 | 0035 | 1 | Pellaea atropurpurea (L.) Link |
| 1 | 2 | 0036 | 1 | Pellaea glabella Mett. ex Kuhn |

ASPLENIACEAE

| | | | | |
|---|---|---|---|---|
| 3 | 3 | 0033A1 | | Asplenium montanum Willd. |
| 3 | 3 | 0029 | 1 | Asplenium pinnatifidum Nutt. |
| 3 | 3 | 0030 | 1 | Asplenium platyneuron (L.) Oakes ex D.C. Eat. |
| 3 | 3 | 0028 | 1 | Asplenium rhizophyllum L. |
| 3 | 3 | 0033 | 1 | Asplenium ruta-muraria L. var. cryptolepis (Fern.) Wherry |
| 3 | 3 | 0032 | 1 | Asplenium trichomanes L. |
| 3 | 3 | 0031 | 1 | Asplenium X ebenoides R.R. Scott |
| 3 | 3 | 0032A1 | | Asplenium X herb-wagneri Taylor & Mohlenbrock |
| 3 | 4 | 0027 | 1 | Athyrium filix-femina (L.) Roth var. angustum (Willd.) Lawson |
| 3 | 4 | 0026 | 1 | Athyrium filix-femina (L.) Roth var. asplenioides (Michx.) Farw. |
| 3 | 4 | 0027A | 1 | Athyrium filix-femina (L.) Roth var. cyclosorum Rupr. |
| 3 | 4 | 0010 | 1 | Cystopteris bulbifera (L.) Bernh. |
| 3 | 4 | 0011 | 1 | Cystopteris fragilis (L.) Bernh. |
| 3 | 4 | 0011A1 | | Cystopteris tennesseensis Shaver |
| 3 | 5 | 0025 | 1 | Deparia acrostichoides (Sw.) M. Kato ined. |
| 3 | 5 | 0024 | 1 | Diplazium pycnocarpon (Spreng.) Broun |
| 3 | 5 | 0021A1 | | Dryopteris celsa (Wm. Palmer) Knowlt., Palmer & Pollard ex Small |
| 3 | 5 | 0019A | 1 | Dryopteris clintoniana (D.C. Eat.) Dowell |
| 3 | 5 | 0019 | 1 | Dryopteris cristata (L.) Gray |
| 3 | 5 | 0018 | 1 | Dryopteris goldiana (Hook.) Gray |
| 3 | 5 | 0018A4 | | Dryopteris intermedia (Willd.) Gray |

1

ASPLENIACEAE     CONT.

| | | | | |
|---|---|---|---|---|
| 3 | 5 | 0017 | 1 | Dryopteris marginalis (L.) Gray |
| 3 | 5 | 0020 | 1 | Dryopteris spinulosa (O.F. Muell.) Watt |
| 3 | 5 | 0021 | 1 | Dryopteris X boottii (Tuckerman) Underwood |
| 3 | 6 | 0012 | 1 | Matteuccia struthiopteris (L.) Todaro |
| 3 | 6 | 0013 | 1 | Onoclea sensibilis L. |
| 3 | 7 | 0022 | 1 | Polystichum acrostichoides (Michx.) Schott |
| 3 | 7 | 0014 | 1 | Thelypteris hexagonoptera (Michx.) Weatherby |
| 3 | 7 | 0015 | 1 | Thelypteris noveboracensis (L.) Nieuwl. |
| 3 | 7 | 0016 | 1 | Thelypteris palustris  Schott var. pubescens (Lawson) Fern. |
| 3 | 8 | 0009 | 1 | Woodsia obtusa (Spreng.) Torr. |

AZOLLACEAE

| | | | | |
|---|---|---|---|---|
| 9 | 9 | 0042 | 1 | Azolla caroliniana Willd. |

BLECHNACEAE

| | | | | |
|---|---|---|---|---|
| 9 | 9 | 0034 | 1 | Woodwardia virginica (L.) Sm. |

DENNSTAEDTIACEAE

| | | | | |
|---|---|---|---|---|
| 10 | 10 | 0023 | 1 | Dennstaedtia punctilobula (Michx.) T. Moore |
| 10 | 10 | 0039 | 1 | Pteridium aquilinum (L.) Kuhn var. latiusculum (Desv.) Underwood ex Heller |

EQUISETACEAE

| | | | | |
|---|---|---|---|---|
| 10 | 10 | 0043 | 1 | Equisetum arvense L. |
| 10 | 10 | 0050 | 1 | Equisetum fluviatile L. |
| 10 | 10 | 0047 | 1 | Equisetum hyemale L. var. affine (Engelm.) A.A. Eat. |
| 10 | 10 | 0048 | 1 | Equisetum laevigatum A. Braun |
| 10 | 10 | 0045 | 1 | Equisetum variegatum Schleich. ex Weber & C. Mohr. |
| 10 | 10 | 0046 | 1 | Equisetum X nelsonii (A.A. Eat.) Schaffn. |
| 10 | 10 | 0044 | 1 | Equisetum X trachyodon A. Braun |

## HYMENOPHYLLACEAE

11   12   0039A3   Trichomanes boschianum Sturm

## ISOETACEAE

12   12   0058   1   Isoetes engelmannii A. Braun

## LYCOPODIACEAE

12   13   0055A      Lycopodium clavatum L.
12   13   0055    1  Lycopodium digitatum A. Braun
12   13   0053    1  Lycopodium inundatum L.
12   13   0052    1  Lycopodium lucidulum Michx.
12   13   0054    1  Lycopodium obscurum L.
12   13   0051    1  Lycopodium selago L. var. selago
12   13   0051A      Lycopodium tristachyum Pursh

## OPHIOGLOSSACEAE

15   15   0004B   1   Botrychium biternatum (Savaiter)
                      Underwood
15   15   0004    1   Botrychium dissectum Spreng.
15   15   0003A4      Botrychium matricariifolium (A. Braun ex
                      Doll) A. Braun ex Koch
15   15   0003    1   Botrychium multifidum (Gmel.) Rupr. var.
                      intermedium (D.C. Eat.) Farw.
15   15   0004C   1   Botrychium oneidense (Gilbert) House
15   15   0002    1   Botrychium simplex E. Hitchc.
15   15   0005    1   Botrychium virginianum (L.) Sw.
15   15   0001A   1   Ophioglossum engelmannii Prantl
15   15   0001    1   Ophioglossum vulgatum L.

## OSMUNDACEAE

15   15   0008    1   Osmunda cinnamomea L. f. auriculata
                      (Hopkins) Kittredge
15   15   0007    1   Osmunda claytoniana L.
15   15   0006    1   Osmunda regalis L. var. spectabilis
                      (Willd.) Gray

## POLYPODIACEAE

16   16   0041    1   Polypodium polypodioides (L.) Watt var.
                      michauxianum Weatherby
16   16   0040    1   Polypodium virginianum L.

3

## SELAGINELLACEAE

| | | | |
|---|---|---|---|
| 17 | 17 | 0056 | 1 Selaginella apoda (L.) Fern. |
| 17 | 17 | 0057 | 1 Selaginella rupestris (L.) Spring |

## CUPRESSACEAE

| | | | |
|---|---|---|---|
| 18 | 18 | 0067 | 1 Juniperus communis L. |
| 18 | 18 | 0068 | 1 Juniperus virginiana L. |
| 18 | 18 | 0066 | 1 Thuja occidentalis L. |

## PINACEAE

| | | | |
|---|---|---|---|
| 19 | 19 | 0063 | 1 Larix laricina (Du Roi) K. Koch |
| 19 | 19 | 0061 | 1 Pinus banksiana Lamb. |
| 19 | 19 | 0060 | 1 Pinus strobus L. |
| 19 | 19 | 0062 | 1 Pinus virginiana P. Mill. |
| 19 | 20 | 0064 | 1 Tsuga canadensis (L.) Carr. |

## TAXACEAE

| | | | |
|---|---|---|---|
| 20 | 20 | 0059 | 1 Taxus canadensis Marsh. |

## TAXODIACEAE

| | | | |
|---|---|---|---|
| 20 | 20 | 0065 | 1 Taxodium distichum (L.) L.C. Rich. |

## ACANTHACEAE

| | | | |
|---|---|---|---|
| 21 | 21 | 1895 | 1 Dicliptera brachiata (Pursh) Spreng. |
| 21 | 21 | 1896 | 1 Justicia americana (L.) Vahl |
| 21 | 22 | 1892 | 1 Ruellia caroliniensis (J.F. Gmel.) Steud. ssp. caroliniensis |
| 21 | 22 | 1892A | Ruellia humilis Nutt. |
| 21 | 22 | 1894 | 1 Ruellia strepens L. |

## ACERACEAE

| | | | |
|---|---|---|---|
| 23 | 23 | 1362 | 1 Acer negundo L. |
| 23 | 23 | 1363 | 1 Acer negundo L. var. violaceum J. Miller |
| 23 | 23 | 1366 | 1 Acer nigrum Michx. f. |
| 23 | 23 | 1366D | Acer platanoides L. |
| 23 | 23 | 1365 | 1 Acer rubrum L. |

4

ACERACEAE          CONT.

23   23   1365A  1  Acer rubrum L. var. drummondii (Hook. &
                     Arnott Ex. Nutt.) Sarg.
23   23   1364   1  Acer saccharinum L.
23   23   1367   1  Acer saccharum Marsh.
23   23   1368   1  Acer saccharum Marsh. f. walpolei
                     (Rehd.) Deam
23   23   1368A  1  Acer saccharum Marsh. var. schneckii
                     Rehd.


AGAVACEAE

24   25   0675   1  Manfreda virginica (L.) Rose
24   25   0649C  2  Yucca filamentosa L.
24   25   0649D     Yucca flaccida Haw.


AIZOACEAE

26   26   0891   2  Mollugo verticillata L.


ALISMACEAE

26   26   0105A  1  Alisma plantago-aquatica L. var.
                     americana Schultes & Schultes
26   26   0105   1  Alisma plantago-aquatica L. var.
                     parviflorum (Pursh) Torr.
26   26   0107   1  Echinodorus cordifolius (L.) Griseb.
26   27   0113   1  Sagittaria australis (J.G. Sm.) Small
26   27   0112   1  Sagittaria brevirostra Mackenzie & Bush
26   27   0108   1  Sagittaria calycina Engelm.
26   27   0111   1  Sagittaria cuneata Sheldon
26   27   0115   1  Sagittaria graminea Michx.
26   27   0109   1  Sagittaria latifolia Willd.
26   27   0110   1  Sagittaria latifolia Willd. var. obtusa
                     (Muhl. ex Willd.) Wieg.
26   27   0114   1  Sagittaria rigida Pursh


AMARANTHACEAE

27   28   0883   1  Amaranthus albus L.
27   28   0882   1  Amaranthus blitoides S. Wats.
27   28   0878A  2  Amaranthus cruentus L.
27   28   0879   2  Amaranthus hybridus L.
27   28   0879A     Amaranthus powellii S. Wats.
27   28   0880   2  Amaranthus retroflexus L.
27   28   0884   1  Amaranthus rudis Sauer

5

## AMARANTHACEAE    CONT.

| | | | | |
|---|---|---|---|---|
| 27 | 28 | 0881 | 2 | Amaranthus spinosus L. |
| 27 | 28 | 0885 | 1 | Amaranthus tuberculatus (Moq.) Sauer |
| 27 | 29 | 0887A | 2 | Froelichia floridana (Nutt.) Moq. var. campestris (Small) Fern. |
| 27 | 29 | 0887 | 2 | Froelichia gracilis (Hook.) Moq. |
| 27 | 29 | 0888 | 1 | Iresine rhizomatosa Standl. |

## ANACARDIACEAE

| | | | | |
|---|---|---|---|---|
| 30 | 30 | 1353 | 1 | Rhus aromatica Ait. |
| 30 | 30 | 1353A | 1 | Rhus aromatica Ait. var. arenaria ('Greene) Fern. |
| 30 | 30 | 1347 | 1 | Rhus copallina L. |
| 30 | 30 | 1349 | 1 | Rhus glabra L. |
| 30 | 30 | 1350 | 1 | Rhus typhina L. |
| 30 | 30 | 1352 | 1 | Toxicodendron radicans (L.) Kuntze ssp. radicans |
| 30 | 30 | 1348 | 1 | Toxicodendron vernix (L.) Kuntze |

## ANNONACEAE

| | | | | |
|---|---|---|---|---|
| 31 | 31 | 0987 | 1 | Asimina triloba (L.) Dunal |

## APIACEAE

| | | | | |
|---|---|---|---|---|
| 31 | 31 | 1540A | | Aethusa cynapium L. |
| 31 | 31 | 1540B | | Anethum graveolens L. |
| 31 | 31 | 1546 | 1 | Angelica atropurpurea L. |
| 31 | 31 | 1545 | 1 | Angelica venenosa (Greenway) Fern. |
| 31 | 32 | 1531 | 2 | Bupleurum rotundifolium L. |
| 31 | 32 | 1537 | 2 | Carum carvi L. |
| 31 | 32 | 1521 | 1 | Chaerophyllum procumbens (L.) Crantz var. procumbens |
| 31 | 32 | 1522 | 1 | Chaerophyllum procumbens (L.) Crantz var. shortii Torr. & Gray |
| 31 | 32 | 1523 | 1 | Chaerophyllum tainturieri Hook. |
| 31 | 32 | 1534 | 1 | Cicuta bulbifera L. |
| 31 | 32 | 1535 | 1 | Cicuta maculata L. |
| 31 | 32 | 1544 | 1 | Conioselinum chinense (L.) B.S.P. |
| 31 | 32 | 1530 | 2 | Conium maculatum L. |
| 31 | 32 | 1536 | 1 | Cryptotaenia canadensis (L.) DC. |
| 31 | 33 | 1551 | 2 | Daucus carota L. |
| 31 | 33 | 1529 | 1 | Erigenia bulbosa (Michx.) Nutt. |
| 31 | 33 | 1519 | 1 | Eryngium yuccifolium Michx. |
| 31 | 33 | 1550 | 1 | Heracleum lanatum Michx. |
| 31 | 34 | 1514 | 1 | Hydrocotyle americana L. |

## APIACEAE          CONT.

| | | | | |
|---|---|---|---|---|
| 31 | 34 | 1513 | 1 | Hydrocotyle umbellata L. |
| 31 | 34 | 1541 | 1 | Ligusticum canadense (L.) Britt. |
| 31 | 36 | 1524 | 1 | Osmorhiza claytonii (Michx.) C.B. Clarke |
| 31 | 36 | 1525 | 1 | Osmorhiza longistylis (Torr.) DC. |
| 31 | 36 | 1548 | 1 | Oxypolis rigidior (L.) Raf. |
| 31 | 36 | 1549 | 2 | Pastinaca sativa L. |
| 31 | 36 | 1539 | 1 | Perideridia americana (Nutt. ex DC.) Reichenb. |
| 31 | 36 | 1547 | 1 | Polytaenia nuttallii DC. |
| 31 | 37 | 1517 | 1 | Sanicula canadensis L. |
| 31 | 37 | 1516 | 1 | Sanicula gregaria Bickn. |
| 31 | 37 | 1515 | 1 | Sanicula marilandica L. |
| 31 | 37 | 1518 | 1 | Sanicula trifoliata Bickn. |
| 31 | 37 | 1540 | 1 | Sium suave Walt. |
| 31 | 37 | 1531A | | Spermolepis inermis (Nutt.) Mathias & Constance |
| 31 | 37 | 1538 | 1 | Taenidia integerrima (L.) Drude |
| 31 | 37 | 1543 | 1 | Thaspium barbinode (Michx.) Nutt. |
| 31 | 37 | 1542 | 1 | Thaspium trifoliatum (L.) Gray var. flavum Blake |
| 31 | 38 | 1528 | 2 | Torilis japonica (Houtt.) DC. |
| 31 | 38 | 1533 | 1 | Zizia aptera (Gray) Fern. |
| 31 | 38 | 1532 | 1 | Zizia aurea (L.) W.D.J. Koch |

## APOCYNACEAE

| | | | | |
|---|---|---|---|---|
| 38 | 38 | 1632 | 1 | Amsonia tabernaemontana Walt. var. salicifolia (Pursh) Woods. |
| 38 | 38 | 1634 | 1 | Apocynum androsaemifolium L. ssp. androsaemifolium |
| 38 | 38 | 1638 | 1 | Apocynum cannabinum L. var. cannabinum |
| 38 | 38 | 1640 | 1 | Apocynum cannabinum L. var. glaberrimum A. DC. |
| 38 | 38 | 1641 | 1 | Apocynum cannabinum L. var. hypericifolium Gray |
| 38 | 38 | 1635 | 1 | Apocynum X medium Greene |
| 38 | 39 | 1633 | 1 | Trachelospermum difforme (Walt.) Gray |
| 38 | 40 | 1632A | 2 | Vinca minor L. |

## AQUIFOLIACEAEAE

| | | | | |
|---|---|---|---|---|
| 40 | 40 | 1354 | 1 | Ilex decidua Walt. |
| 40 | 40 | 1355 | 1 | Ilex verticillata (L.) Gray |
| 40 | 40 | 1356 | 1 | Nemopanthus mucronata (L.) Trel. |

7

## ARACEAE

| | | | | |
|---|---|---|---|---|
| 40 | 40 | 0570 | 1 | Acorus americanus (Raf.) Raf. |
| 40 | 40 | 0574 | 1 | Arisaema dracontium (L.) Schott |
| 40 | 40 | 0575 | 1 | Arisaema triphyllum (L.) Schott ssp. pusillum (Peck) Huttleston |
| 40 | 40 | 0575A | | Arisaema triphyllum (L.) Schott ssp. stewardsonii (Britt.) Huttleston |
| 40 | 40 | 0576 | 1 | Arisaema triphyllum (L.) Schott ssp. triphyllum |
| 40 | 40 | 0572 | 1 | Calla palustris L. |
| 40 | 41 | 0573 | 1 | Peltandra virginica (L.) Schott ssp. virginica |
| 40 | 41 | 0571 | 1 | Symplocarpus foetidus (L.) Nutt. |

## ARALIACEAE

| | | | | |
|---|---|---|---|---|
| 41 | 41 | 1509 | 1 | Aralia hispida Vent. |
| 41 | 41 | 1507 | 1 | Aralia nudicaulis L. |
| 41 | 41 | 1510 | 1 | Aralia racemosa L. |
| 41 | 41 | 1508 | 1 | Aralia spinosa L. |
| 41 | 42 | 1511 | 1 | Panax quinquefolius L. |
| 41 | 42 | 1512 | 1 | Panax trifolius L. |

## ARISTOLOCHIACEAE

| | | | | |
|---|---|---|---|---|
| 43 | 43 | 0823 | 1 | Aristolochia serpentaria L. |
| 43 | 43 | 0824 | 1 | Aristolochia tomentosa Sims |
| 43 | 44 | 0822 | 1 | Asarum canadense L. var. canadense |
| 43 | 44 | 0821 | 1 | Asarum canadense L. var. reflexum (Bickn.) B.L. Robins. |

## ASCLEPIADACEAE

| | | | | |
|---|---|---|---|---|
| 44 | 44 | 1648 | 1 | Asclepias amplexicaulis Sm. |
| 44 | 44 | 1654 | 1 | Asclepias exaltata L. |
| 44 | 44 | 1643 | 1 | Asclepias hirtella (Pennell) Woods. |
| 44 | 44 | 1650 | 1 | Asclepias incarnata L. ssp. incarnata |
| 44 | 44 | 1647 | 1 | Asclepias meadii Torr. ex Gray |
| 44 | 44 | 1652 | 1 | Asclepias perennis Walt. |
| 44 | 44 | 1656 | 1 | Asclepias purpurascens L. |
| 44 | 44 | 1651 | 1 | Asclepias quadrifolia Jacq. |
| 44 | 44 | 1649 | 1 | Asclepias sullivantii Engelm. ex Gray |
| 44 | 44 | 1655 | 1 | Asclepias syriaca L. |
| 44 | 44 | 1646 | 1 | Asclepias tuberosa L. |
| 44 | 44 | 1653 | 1 | Asclepias variegata L. |
| 44 | 44 | 1645 | 1 | Asclepias verticillata L. |
| 44 | 44 | 1644 | 1 | Asclepias viridiflora Raf. |

ASCLEPIADACEAE    CONT.

44   45   1657   1 Cynanchum laeve (Michx.) Pers.
44   46   1658   1 Matelea gonocarpa (Walt.) Shinners
44   46   1659   1 Matelea obliqua (Jacq.) Woods.


ASTERACEAE

46   47   2170   1 Achillea millefolium L.
46   48   2105   1 Ambrosia artemisiifolia L. var. elatior
                   Descourtils
46   48   2103   2 Ambrosia bidentata Michx.
46   48   2106   2 Ambrosia coronopifolia Torr. & Gray
46   48   2104   1 Ambrosia trifida L.
46   48   2000A    Amphiachyris dracunculoides (DC.) Nutt.
46   48   2086A4   Anaphalis margaritaceae (L.) Benth. &
                   Hook. f. ex C.B. Clarke
46   49   2079   1 Antennaria neglecta Greene var.
                   attenuata (Fern.) Cronq.
46   49   2078   1 Antennaria neglecta Greene var. neglecta
46   49   2083   1 Antennaria plantaginifolia (L.) Richards
                   var. ambigens (Greene) Cronq.
46   49   2081   1 Antennaria plantaginifolia (L.) Richards
                   var. arnoglossa (Greene) Cronq.
46   49   2082   1 Antennaria plantaginifolia (L.) Richards
                   var. plantaginifolia
46   49   2080   1 Antennaria solitaria Rydb.
46   51   2168   2 Anthemis arvensis L.
46   51   2167   2 Anthemis cotula L.
46   51   2192A4   Arctium lappa L.
46   51   2192   2 Arctium minus Bernh.
46   52   2175   2 Artemisia absinthium L.
46   52   2177   2 Artemisia annua L.
46   52   2176   1 Artemisia biennis Willd.
46   52   2178   1 Artemisia campestris L. ssp. caudata
                   (Michx.) Hall & Clements
46   52   2176A    Artemisia carruthii Wood ex Carruthers
46   52   2179   2 Artemisia ludoviciana Nutt. ssp.
                   ludoviciana
46   52   2179A    Artemisia vulgaris L.
46   53   2035   1 Aster azureus Lindl.
46   53   2052   1 Aster borealis (Torr. & Gray) Prov.
46   53   2035A    Aster brachyactis Blake
46   53   2037   1 Aster cordifolius L.
46   53   2040   1 Aster drummondii Lindl.
46   53   2067   1 Aster dumosus L.
46   53   2058   1 Aster ericoides L. ssp. ericoides var.
                   ericoides
46   53   2049   1 Aster firmus Nees
46   53   2034   1 Aster furcatus Burgess

9

| 46 | 53 | 2047 | 1 | Aster laevis L. |
|----|----|------|---|-----------------|
| 46 | 53 | 2066 | 1 | Aster lateriflorus (L.) Britt. var. angustifolius Wieg. |
| 46 | 53 | 2065 | 1 | Aster lateriflorus (L.) Britt. var. lateriflorus |
| 46 | 53 | 2056 | 1 | Aster linariifolius L. |
| 46 | 53 | 2048 | 1 | Aster longifolius Lam. |
| 46 | 53 | 2033 | 1 | Aster macrophyllus L. |
| 46 | 53 | 2033A | 1 | Aster macrophyllus L. var. ianthinus (Burgess) Fern. |
| 46 | 53 | 2033B | 1 | Aster macrophyllus L. var. pinguifolius Burgess |
| 46 | 53 | 2033C | 1 | Aster macrophyllus L. var. velutinus Burgess |
| 46 | 53 | 2042 | 1 | Aster novae-angliae L. |
| 46 | 53 | 2044 | 1 | Aster oblongifolius Nutt. |
| 46 | 53 | 2062 | 1 | Aster ontarionis Wieg. |
| 46 | 53 | 2043 | 1 | Aster patens Ait. |
| 46 | 53 | 2061B | 1 | Aster pilosus Willd. var. demotus Blake |
| 46 | 53 | 2061 | 1 | Aster pilosus Willd. var. pilosus |
| 46 | 53 | 2051 | 1 | Aster praealtus Poir. var. angustior Wieg. |
| 46 | 53 | 2050 | 1 | Aster praealtus Poir. var. praealtus |
| 46 | 53 | 2046 | 1 | Aster prenanthoides Muhl. |
| 46 | 53 | 2057 | 1 | Aster puniceus L. |
| 46 | 53 | 2060 | 1 | Aster sericeus Vent. |
| 46 | 53 | 2036 | 1 | Aster shortii Lindl. |
| 46 | 53 | 2053 | 1 | Aster simplex Willd. var. ramosissimus (Torr. & Gray) Cronq. |
| 46 | 53 | 2054 | 1 | Aster simplex Willd. var. simplex |
| 46 | 53 | 2076 | 1 | Aster solidagineus Michx. |
| 46 | 53 | 2067A | 1 | Aster tradescantii L. |
| 46 | 53 | 2063 | 1 | Aster umbellatus P. Mill. |
| 46 | 53 | 2041 | 1 | Aster undulatus L. |
| 46 | 53 | 2039 | 1 | Aster urophyllus Lindl. |
| 46 | 53 | 2068 | 1 | Aster vimineus Lam. |
| 46 | 53 | 2068A | 1 | Aster vimineus Lam. var. subdumosus Wieg. |
| 46 | 53 | 2055 | 1 | Aster X interior Wieg. |
| 46 | 53 | 2038 | 1 | Aster X sagittifolius Wedemeyer ex Willd. |
| 46 | 58 | 2173 | 2 | Balsamita major Desf. |
| 46 | 59 | 2075A4 |  | Bellis perennis L. |
| 46 | 59 | 2155 | 1 | Bidens aristosa (Michx.) Britt. var. aristosa |
| 46 | 59 | 2159A | 1 | Bidens aristosa (Michx.) Britt. var. retrorsa (Sherff) Wunderlin |
| 46 | 59 | 2152 | 1 | Bidens bipinnata L. |
| 46 | 59 | 2148 | 1 | Bidens cernua L. |
| 46 | 59 | 2150 | 1 | Bidens connata Muhl. ex Willd. |
| 46 | 59 | 2154 | 1 | Bidens coronata (L.) Britt. |

```
46   59   2153    1 Bidens discoidea (Torr. & Gray) Britt.
46   59   2158    1 Bidens frondosa L.
46   59   2149    1 Bidens tripartita L.
46   59   2159    1 Bidens vulgata Greene
46   60   2032      Boltonia asteroides (L.) L'Her
46   60   2032A     Boltonia asteroides (L.) L'Her. var.
                    recognita (Fern. & Grisc.) Cronq.
46   61   1993    1 Brickellia eupatorioides (L.) Shinners
46   61   1994    1 Brickellia eupatorioides (L.) Shinners
                    var. corymbulosa (Torr. & Gray) Shinners
46   61   2182    1 Cacalia atriplicifolia L.
46   61   2183    1 Cacalia muhlenbergii (Schultz-Bip.) Fern.
46   61   2184    1 Cacalia plantaginea (Raf.) Shinners
46   61   2181    1 Cacalia suaveolens L.
46   62   2193    2 Carduus nutans L.
46   62   2202A     Centaurea cyanus L.
46   62   2202A4    Centaurea diffusa Lam.
46   62   2202B     Centaurea maculosa Lam.
46   62   2202C     Centaurea moschata L.
46   63   2169    2 Chamaemelum nobilis (L.) All.
46   65   2203    2 Cichorium intybus L.
46   65   2201    1 Cirsium altissimum (L.) Spreng.
46   65   2196    2 Cirsium arvense (L.) Scop.
46   65   2196B   2 Cirsium arvense (L.) Scop. var.
                    integrifolium Wimmer & Grab.
46   65   2196A   2 Cirsium arvense (L.) Scop. var. mite
                    Wimmer & Grab.
46   65   2196C   2 Cirsium arvense (L.) Scop. var. vestitum
                    Wimmer & Grab.
46   65   2200    1 Cirsium discolor (Muhl. ex Willd.)
                    Spreng.
46   65   2197    1 Cirsium hillii (Canby) Fern.
46   65   2198    1 Cirsium muticum Michx.
46   65   2195    1 Cirsium pitcheri (Torr. ex Eat.) Torr. &
                    Gray
46   65   2199    1 Cirsium virginianum (L.) Michx.
46   65   2194    2 Cirsium vulgare (Savi) Tenore
46   67   2069    1 Conyza canadensis (L.) Cronq.
46   67   2070    1 Conyza canadensis (L.) Cronq. var.
                    pusilla (Nutt.) Cronq.
46   67   2071    1 Conyza ramosissima Cronq.
46   67   2145A   2 Coreopsis grandiflora Hogg ex Sweet
46   67   2143    1 Coreopsis lanceolata L.
46   67   2145    1 Coreopsis palmata Nutt.
46   67   2145A4    Coreopsis tinctoria Nutt.
46   67   2146    1 Coreopsis tripteris L.
46   68   2229    2 Crepis capillaris (L.) Wallr.
46   68   2230    2 Crepis pulchra L.
46   70   2166    2 Dyssodia papposa (Vent.) A.S. Hitchc.
```

| 46 | 71 | 2122A | 1 | Echinacea angustifolia DC. |
| 46 | 71 | 2122 | 1 | Echinacea pallida Nutt. |
| 46 | 71 | 2121 | 1 | Echinacea purpurea (L.) Moench |
| 46 | 71 | 2111 | 1 | Eclipta prostrata (L.) L. |
| 46 | 71 | 1981 | 1 | Elephantopus carolinianus Raeusch. |
| 46 | 72 | 2180 | 1 | Erechtites hieraciifolia (L.) Raf. ex DC. |
| 46 | 72 | 2075 | 1 | Erigeron annuus (L.) Pers. |
| 46 | 72 | 2073 | 1 | Erigeron philadelphicus L. |
| 46 | 72 | 2072 | 1 | Erigeron pulchellus Michx. |
| 46 | 72 | 2074 | 1 | Erigeron strigosus Muhl. ex Willd. var. strigosus |
| 46 | 75 | 1986 | 1 | Eupatorium altissimum L. |
| 46 | 75 | 1991 | 1 | Eupatorium coelestinum L. |
| 46 | 75 | 1983 | 1 | Eupatorium fistulosum Barratt |
| 46 | 75 | 1990 | 1 | Eupatorium incarnatum Walt. |
| 46 | 75 | 1982 | 1 | Eupatorium maculatum L. |
| 46 | 75 | 1988 | 1 | Eupatorium perfoliatum L. |
| 46 | 75 | 1984 | 1 | Eupatorium purpureum L. |
| 46 | 75 | 1989 | 1 | Eupatorium rugosum Houtt. |
| 46 | 75 | 1985 | 1 | Eupatorium serotinum Michx. |
| 46 | 75 | 1987 | 1 | Eupatorium sessilifolium L. |
| 46 | 75 | 1988B | 1 | Eupatorium x polyneuron (F.J. Herm.) Wunderlin |
| 46 | 76 | 2029 | 1 | Euthamia graminifolia (L.) Nutt. ex Cass. var. nuttallii (Greene) W. Stone |
| 46 | 76 | 2030 | 1 | Euthamia gymnospermoides Greene |
| 46 | 76 | 2031 | 1 | Euthamia tenuifolia (Pursh) Greene |
| 46 | 77 | 2161A4 | | Gaillardia pulchella Foug. |
| 46 | 77 | 2161C | | Galinsoga parviflora Cav. |
| 46 | 77 | 2161 | 2 | Galinsoga quadriradiata Ruiz & Pavon |
| 46 | 78 | 2087 | 1 | Gnaphalium obtusifolium L. |
| 46 | 78 | 2090 | 1 | Gnaphalium purpureum L. |
| 46 | 78 | 2089 | 1 | Gnaphalium uliginosum L. |
| 46 | 78 | 2088 | 1 | Gnaphalium viscosum H.B.K. |
| 46 | 78 | 2000 | 2 | Grindelia squarrosa (Pursh) Dunal |
| 46 | 78 | 2000B | 2 | Grindelia squarrosa (Pursh) Dunal var. serrulata (Rydb.) Steyermark |
| 46 | 80 | 2163 | 2 | Helenium amarum (Raf.) H. Rock |
| 46 | 80 | 2164 | 1 | Helenium autumnale L. |
| 46 | 80 | 2165 | 1 | Helenium flexuosum Raf. |
| 46 | 81 | 2125 | 2 | Helianthus angustifolius L. |
| 46 | 81 | 2127 | 2 | Helianthus annuus L. |
| 46 | 81 | 2138 | 1 | Helianthus decapetalus L. |
| 46 | 81 | 2131 | 1 | Helianthus divaricatus L. |
| 46 | 81 | 2136 | 1 | Helianthus giganteus L. |
| 46 | 81 | 2134 | 1 | Helianthus grosseserratus Martens |
| 46 | 81 | 2137 | 1 | Helianthus hirsutus Raf. |
| 46 | 81 | 2135 | 2 | Helianthus maximilianii Schrad. |
| 46 | 81 | 2130 | 1 | Helianthus microcephalus Torr. & Gray |

ASTERACEAE      CONT.

```
46   81   2132    1 Helianthus mollis Lam.
46   81   2128    1 Helianthus occidentalis Riddell
46   81   2126    2 Helianthus petiolaris Nutt.
46   81   2129    1 Helianthus rigidus (Cass.) Desf.
46   81   2140    1 Helianthus strumosus L.
46   81   2139    1 Helianthus tuberosus L.
46   81   2133    1 Helianthus X doronicoides Lam.
46   82   2110    1 Heliopsis helianthoides (L.) Sweet
46   83   2001A     Heterotheca camporum (Greene) Shinners
46   83   2001    1 Heterotheca villosa (Pursh) Shinners
46   84   2237    2 Hieracium aurantiacum L.
46   84   2238A4    Hieracium caespitosum Dumort.
46   84   2238    1 Hieracium canadense Michx.
46   84   2239    1 Hieracium gronovii L.
46   84   2240    1 Hieracium longipilum Torr.
46   84   2242    1 Hieracium paniculatum L.
46   84   2243    1 Hieracium scabrum Michx.
46   84   2241    1 Hieracium venosum L.
46   85   2162    2 Hymenopappus scabiosaeus L'Her.
46   86   2207A     Hypochoeris radicata L.
46   86   2091    1 Inula helenium L.
46   86   2101    1 Iva annua L. var. annua
46   86   2102    1 Iva xanthifolia Nutt.
46   86   2207    1 Krigia biflora  (Walt.) Blake
46   86   2204    1 Krigia caespitosa (Raf.) Chambers
46   86   2205    1 Krigia dandelion (L.) Nutt.
46   86   2206    1 Krigia virginica (L.) Willd.
46   87   2226    1 Lactuca biennis (Moench) Fern.
46   87   2220    1 Lactuca canadensis L. var. canadensis
46   87   2221    1 Lactuca canadensis L. var. latifolia
                     Kuntze
46   87   2223    1 Lactuca canadensis L. var. obovata Wieg.
46   87   2225    1 Lactuca floridana (L.) Gaertn.
46   87   2224    1 Lactuca floridana (L.) Gaertn. var.
                     villosa (Jacq.) Cronq.
46   87   2218    1 Lactuca ludoviciana (Nutt.) Riddell
46   87   2219    2 Lactuca saligna L.
46   87   2217    2 Lactuca sativa L.
46   87   2216    2 Lactuca serriola L.
46   87   2207B     Lapsana communis L.
46   88   2172    2 Leucanthemum vulgare Lam.
46   88   1997A     Liatris aspera Michx.
46   88   1996    1 Liatris cylindracea Michx.
46   88   1997    1 Liatris pycnostachya Michx. var.
                     pycnostachya
46   88   1999    1 Liatris scariosa (L.) Willd.
46   88   1998    1 Liatris spicata (L.) Willd.
46   88   1995    1 Liatris squarrosa (L.) Michx.
46   92   2161A   2 Madia capitata Nutt.
```

| 46 | 92  | 2171A |   | Matricaria chamomilla L. |
|----|-----|-------|---|--------------------------|
| 46 | 92  | 2171  | 2 | Matricaria matricarioides (Less.) Porter |
| 46 | 93  | 2160  | 1 | Megalodonta beckii (Torr. ex Spreng.) Greene |
| 46 | 93  | 1992  | 1 | Mikania scandens (L.) Willd. |
| 46 | 94  | 2202  | 2 | Onopordum acanthium L. |
| 46 | 94  | 2100  | 1 | Parthenium integrifolium L. |
| 46 | 96  | 2077  | 1 | Pluchea camphorata (L.) DC. |
| 46 | 97  | 2092  | 1 | Polymnia canadensis L. |
| 46 | 97  | 2093  | 1 | Polymnia uvedalia L. |
| 46 | 97  | 2233  | 1 | Prenanthes alba L. |
| 46 | 97  | 2231  | 1 | Prenanthes altissima L. |
| 46 | 97  | 2235  | 1 | Prenanthes aspera Michx. |
| 46 | 97  | 2236  | 1 | Prenanthes crepidinea Michx. |
| 46 | 97  | 2234  | 1 | Prenanthes racemosa Michx. |
| 46 | 97  | 2232  | 1 | Prenanthes trifoliolata (Cass.) Fern. |
| 46 | 98  | 2228  | 1 | Pyrrhopappus carolinianus (Walt.) DC. |
| 46 | 98  | 2124  | 2 | Ratibida columnifera (Nutt.) Woot. & Standl. |
| 46 | 98  | 2123  | 1 | Ratibida pinnata (Vent.) Barnh. |
| 46 | 98  | 2120  | 1 | Rudbeckia fulgida Ait. var. deamii (Blake) Perdue |
| 46 | 98  | 2116  | 1 | Rudbeckia fulgida Ait. var. fulgida |
| 46 | 98  | 2119  | 1 | Rudbeckia fulgida Ait. var. palustris (Eggert) Perdue |
| 46 | 98  | 2118  | 1 | Rudbeckia fulgida Ait. var. sullivantii (C.L. Boynt. & Beadle)    Cronq. |
| 46 | 98  | 2117  | 1 | Rudbeckia fulgida Ait. var. umbrosa (C.L. Boynt. & Beadle)    Cronq. |
| 46 | 98  | 2112  | 1 | Rudbeckia hirta L. |
| 46 | 98  | 2114  | 1 | Rudbeckia laciniata L. |
| 46 | 98  | 2113  | 1 | Rudbeckia subtomentosa Pursh |
| 46 | 98  | 2115  | 1 | Rudbeckia triloba L. |
| 46 | 99  | 2189  | 1 | Senecio aureus L. |
| 46 | 99  | 2186  | 1 | Senecio glabellus Poir. |
| 46 | 99  | 2188  | 1 | Senecio obovatus Muhl. ex Willd. |
| 46 | 99  | 2191  | 1 | Senecio pauperculus Michx. |
| 46 | 99  | 2187  | 1 | Senecio plattensis Nutt. |
| 46 | 99  | 2185  | 2 | Senecio vulgaris L. |
| 46 | 101 | 2097  | 1 | Silphium asteriscus L. ssp. trifoliatum (Ell.) Weber & T.R. Fisher ined. |
| 46 | 101 | 2098  | 1 | Silphium integrifolium Michx. |
| 46 | 101 | 2099  | 1 | Silphium integrifolium Michx. var. deamii Perry |
| 46 | 101 | 2096  | 1 | Silphium laciniatum L. |
| 46 | 101 | 2096A | 1 | Silphium laciniatum L. var. robinsonii Perry |
| 46 | 101 | 2095  | 1 | Silphium perfoliatum L. |
| 46 | 101 | 2094  | 1 | Silphium terebinthinaceum Jacq. |

ASTERACEAE        CONT.

```
46  102  2004   1  Solidago bicolor L.
46  102  2003   1  Solidago buckleyi Torr. & Gray
46  102  2007   1  Solidago caesia L.
46  102  2011   1  Solidago canadensis L. var. canadensis
46  102  2012   1  Solidago canadensis L. var.
                   gilvocanescens Rydb.
46  102  2016   1  Solidago canadensis L. var. scabra
                   (Muhl.) Torr. & Gray
46  102  2009   1  Solidago deamii Fern.
46  102  2006   1  Solidago erecta Pursh
46  102  2008   1  Solidago flexicaulis L.
46  102  2014   1  Solidago gigantea Ait. var. gigantea
46  102  2015   1  Solidago gigantea Ait. var. serotina
                   (Ait.) Cronq.
46  102  2005   1  Solidago hispida Muhl.
46  102  2013A  1  Solidago juncea Ait.
46  102  2013   1  Solidago missouriensis Nutt. var.
                   fasciculata Holz.
46  102  2018   1  Solidago nemoralis Ait. var.
                   longipetiolata (Mackenzie & Bush) Palmer & Steyermark
46  102  2017   1  Solidago nemoralis Ait. var. nemoralis
46  102  2027   1  Solidago ohioensis Riddell
46  102  2019   1  Solidago patula Muhl.
46  102  2064   1  Solidago ptarmicoides (Nees) Boivin
46  102  2028   1  Solidago riddellii Frank
46  102  2026   1  Solidago rigida L.
46  102  2022   1  Solidago rugosa Ait. var. aspera (Ait.)
                   Fern.
46  102  2021   1  Solidago rugosa Ait. var. rugosa
46  102  2010   1  Solidago spathulata DC. ssp. spathulata
                   var. gillmanii (Gray) Cronq.
46  102  2025   1  Solidago speciosa Nutt.
46  102  2024   1  Solidago sphacelata Raf.
46  102  2002   1  Solidago squarrosa Muhl.
46  102  2023   1  Solidago uliginosa Nutt.
46  102  2020   1  Solidago ulmifolia Muhl.
46  102  2024A  1  Solidago X ovata Friesner
46  104  2212   2  Sonchus arvensis L. ssp. arvensis
46  104  2214A     Sonchus arvensis L. ssp. uliginosus
                   (Bieb.) Nyman
46  104  2215   2  Sonchus asper (L.) Hill
46  104  2214   2  Sonchus oleraceus L.
46  105  2174   2  Tanacetum vulgare L.
46  105  2211   2  Taraxacum laevigatum (Willd.) DC.
46  105  2210   2  Taraxacum palustre (Lyons) Symons
46  107  2208A     Tragopogon dubius Scop. ssp. major
                   (Jacq.) Voll.
46  107  2208   2  Tragopogon porrifolius L.
46  107  2209   2  Tragopogon pratensis L. ssp. pratensis
```

## ASTERACEAE     CONT.

| | | | |
|---|---|---|---|
| 46 | 107 | 2181A5 | Tussilago farfara L. |
| 46 | 107 | 2141 1 | Verbesina alternifolia (L.) Britt. |
| 46 | 107 | 2142 1 | Verbesina helianthoides Michx. |
| 46 | 108 | 1979 1 | Vernonia fasciculata Michx. |
| 46 | 108 | 1978 1 | Vernonia gigantea (Walt.) Trel. ex Branner & Coville |
| 46 | 108 | 1980 1 | Vernonia missurica Raf. |
| 46 | 109 | 2107 2 | Xanthium spinosum L. |
| 46 | 109 | 2107A | Xanthium strumarium L. |
| 46 | 109 | 2108 1 | Xanthium strumarium L. var. canadense (P. Mill.) Torr. & Gray |

## BALSAMINACEAE

| | | | |
|---|---|---|---|
| 110 | 110 | 1371 1 | Impatiens capensis Meerb. |
| 110 | 110 | 1372 1 | Impatiens pallida Nutt. |

## BERBERIDACEAE

| | | | |
|---|---|---|---|
| 110 | 110 | 0981C 1 | Berberis canadensis P. Mill. |
| 110 | 110 | 0981A 2 | Berberis thunbergii DC. |
| 110 | 110 | 0981 1 | Caulophyllum thalictroides (L.) Michx. |
| 110 | 110 | 0980 1 | Jeffersonia diphylla (L.) Pers. |
| 110 | 111 | 0979 1 | Podophyllum peltatum L. |

## BETULACEAE

| | | | |
|---|---|---|---|
| 111 | 111 | 0776A | Alnus glutinosa (L.) Gaertn. |
| 111 | 111 | 0776B | Alnus incana (L.) Moench |
| 111 | 111 | 0777 1 | Alnus incana (L.) Moench ssp. rugosa (Du Roi) Clausen |
| 111 | 111 | 0777A4 | Alnus serrulata (Ait.) Willd. |
| 111 | 111 | 0773A4 | Betula alleghaniensis Britt. var. alleghaniensis |
| 111 | 111 | 0771 1 | Betula alleghaniensis Britt. var. macrolepis (Fern.) Brayshaw |
| 111 | 111 | 0774 1 | Betula nigra L. |
| 111 | 111 | 0773 1 | Betula papyrifera Marsh. |
| 111 | 111 | 0772 1 | Betula populifolia Marsh. |
| 111 | 111 | 0775 1 | Betula pumila L. |
| 111 | 111 | 0775A 1 | Betula pumila L. var. glandulifera Regel |
| 111 | 111 | 0775B 1 | Betula X purpusii Schneid. |
| 111 | 112 | 0767 1 | Carpinus caroliniana Walt. |
| 111 | 112 | 0770 1 | Corylus americana Walt. |
| 111 | 112 | 0768 1 | Ostrya virginiana (P. Mill.) K. Koch |
| 111 | 112 | 0769 1 | Ostrya virginiana (P. Mill.) K. Koch f. glandulosa (Spach) Macbr. |

## BIGNONIACEAE

| | | | | |
|---|---|---|---|---|
| 112 | 112 | 1874 | 1 | Bignonia capreolata L. |
| 112 | 112 | 1875 | 1 | Campsis radicans (L.) Seem. ex Bureau |
| 112 | 112 | 1876 | 2 | Catalpa bignonioides Walt. |
| 112 | 112 | 1877 | 1 | Catalpa speciosa (Warder ex Barney) Engelm. |
| 112 | 113 | 1835A | 2 | Paulownia tomentosa (Thunb.) Sieb. & Zucc. ex Steud. |

## BORAGINACEAE

| | | | | |
|---|---|---|---|---|
| 113 | 114 | 1711 | 2 | Buglossoides arvense (L.) I.M. Johnston |
| 113 | 116 | 1701A | | Cynoglossum boreale Fern. |
| 113 | 116 | 1701 | 1 | Cynoglossum officinale L. |
| 113 | 116 | 1702 | 1 | Cynoglossum virginianum L. |
| 113 | 116 | 1717 | 2 | Echium vulgare L. |
| 113 | 116 | 1704 | 1 | Hackelia virginiana (L.) I.M. Johnston |
| 113 | 116 | 1700 | 2 | Heliotropium indicum L. |
| 113 | 117 | 1703 | 2 | Lappula echinata Gilib. |
| 113 | 117 | 1714 | 1 | Lithospermum canescens (Michx.) Lehm. |
| 113 | 117 | 1715 | 1 | Lithospermum caroliniense (J.F. Gmel.) MacM. |
| 113 | 117 | 1713 | 1 | Lithospermum incisum Lehm. |
| 113 | 117 | 1712 | 1 | Lithospermum latifolium Michx. |
| 113 | 117 | 1710 | 1 | Mertensia virginica (L.) Pers. ex Link |
| 113 | 118 | 1706 | 1 | Myosotis laxa Lehm. |
| 113 | 118 | 1708 | 1 | Myosotis macrosperma Engelm. |
| 113 | 118 | 1705 | 2 | Myosotis scorpioides L. |
| 113 | 118 | 1709 | 2 | Myosotis stricta Link ex Roemer & Schultes |
| 113 | 118 | 1707 | 1 | Myosotis verna Nutt. |
| 113 | 118 | 1716 | 1 | Onosmodium hispidissimum Mackenzie |
| 113 | 120 | 1704A | | Symphytum officinale L. |

## BRASSICACEAE

| | | | | |
|---|---|---|---|---|
| 120 | 120 | 1002A | | Alliaria petiolata (Bieb.) Cavara & Grande |
| 120 | 120 | 1052 | 2 | Alyssum alyssoides (L.) L. |
| 120 | 120 | 1006 | 2 | Arabidopsis thaliana (L.) Heynh. |
| 120 | 120 | 1048 | 1 | Arabis canadensis L. |
| 120 | 120 | 1046 | 1 | Arabis drummondii Gray |
| 120 | 120 | 1045 | 1 | Arabis glabra (L.) Bernh. |
| 120 | 120 | 1040 | 1 | Arabis hirsuta (L.) Scop. var. adpressipilis (M. Hopkins) Rollins |
| 120 | 120 | 1039 | 1 | Arabis hirsuta (L.) Scop. var. pycnocarpa (M. Hopkins) Rollins |

| | | | |
|---|---|---|---|
| 120 | 120 | 1044 | 1 Arabis laevigata (Muhl.)  Poir. |
| 120 | 120 | 1047 | 1 Arabis lyrata L. |
| 120 | 120 | 1041 | 1 Arabis missouriensis Greene var. deamii |
| | | | (M. Hopkins) M. Hopkins |
| 120 | 120 | 1042 | 1 Arabis patens Sullivant |
| 120 | 120 | 1043 | 1 Arabis shortii (Fern.) Gleason var. |
| | | | shortii |
| 120 | 122 | 1021 | 1 Armoracia aquatica (Eat.) Wieg. |
| 120 | 122 | 1020 | I Armoracia rusticana (Lam.) Gaertn., |
| | | | Mey., & Scherb. |
| 120 | 122 | 1013 | 2 Barbarea verna (P. Mill.) Aschers. |
| 120 | 122 | 1012 | 2 Barbarea vulgaris R. Br. |
| 120 | 122 | 1053 | 2 Berteroa incana (L.) DC. |
| 120 | 122 | 1009 | 2 Brassica juncea (L.) Czern. |
| 120 | 122 | 1010A4 | Brassica napus L. |
| 120 | 122 | 1010 | 2 Brassica nigra (L.) W.D.J. Koch |
| 120 | 122 | 1010A | Brassica rapa L. |
| 120 | 122 | 1008 | 2 Brassica rapa L. ssp. olifera DC. |
| 120 | 123 | 1007 | 1 Cakile edentula (Bigelow) Hook. ssp. |
| | | | lacustris (Fern.) Hulten |
| 120 | 123 | 1033 | 2 Camelina microcarpa Andrz. ex DC. |
| 120 | 123 | 1032 | 2 Capsella bursa-pastoris (L.) Medic. |
| 120 | 123 | 1022 | 1 Cardamine bulbosa (Schreb.) B.S.P. |
| 120 | 123 | 1023 | 1 Cardamine douglassii (Torr.) Britt. |
| 120 | 123 | 1026 | 1 Cardamine parviflora L. ssp. parviflora |
| | | | var. arenicola (Britt.) O.E. Schulz |
| 120 | 123 | 1025 | 1 Cardamine pensylvanica Muhl. ex Willd. |
| 120 | 123 | 1024 | 1 Cardamine pratensis L. |
| 120 | 124 | 0998 | 2 Cardaria draba (L.) Desv. |
| 120 | 124 | 1054A | Chorispora tenella (Pallas) DC. |
| 120 | 147 | 1055 | 2 Conringia orientalis (L.) Dumort. |
| 120 | 124 | 1000B | Coronopus didymus (L.) Sm. |
| 120 | 124 | 1029 | 1 Dentaria diphylla Michx. |
| 120 | 124 | 1030 | 1 Dentaria heterophylla Nutt. |
| 120 | 124 | 1027 | 1 Dentaria laciniata Muhl. ex Willd. |
| 120 | 124 | 1028 | 1 Dentaria multifida Muhl. |
| 120 | 125 | 1037 | 1 Descurainia pinnata (Walt.) Britt. ssp. |
| | | | brachycarpa (Richards.) Detling |
| 120 | 125 | 1037A | Descurainia sophia (L.) Webb. ex Prantl |
| 120 | 125 | 1007C | Diplotaxis muralis (L.) DC. |
| 120 | 125 | 1007E | Diplotaxis tenuifolia (L.) DC. |
| 120 | 125 | 1034 | 1 Draba brachycarpa Nutt. ex Torr. & Gray |
| 120 | 125 | 1036 | 1 Draba reptans (Lam.) Fern. |
| 120 | 127 | 1035 | 2 Erophila verna (L.) Chev. ssp. verna |
| 120 | 127 | 1007D | Erucastrum gallicum (Willd.) O.E. Schulz |
| 120 | 127 | 1049 | 1 Erysimum asperum (Nutt.) DC. |
| 120 | 127 | 1051 | 1 Erysimum cheiranthoides L. |
| 120 | 127 | 1051A | Erysimum inconspicuum (S. Wats.) MacM. |
| 120 | 127 | 1050 | 2 Erysimum repandum L. |

## BRASSICACEAE    CONT.

| | | | |
|---|---|---|---|
| 120 | 128 | 1054 | 2 Hesperis matronalis L. |
| 120 | 128 | 1014 | 1 Iodanthus pinnatifidus (Michx.) Steud. |
| 120 | 128 | 1031 | 1 Leavenworthia uniflora (Michx.) Britt. |
| 120 | 128 | 0997 | 2 Lepidium campestre (L.) R. Br. |
| 120 | 128 | 1000 | 2 Lepidium densiflorum Schrad. var. densiflorum |
| 120 | 128 | 0999 | 1 Lepidium virginicum L. var. virginicum |
| 120 | 129 | 1032A1 | Lesquerella globosa (Desv.) S. Wats. |
| 120 | 129 | 1932A4 | Lesquerella gracilis (Hook.) S. Wats. |
| 120 | 130 | 1019 | 2 Nasturtium officinale R. Br. |
| 120 | 130 | 1033A | Neslia paniculata (L.) Desv. |
| 120 | 131 | 1011A | 2 Raphanus raphanistrum L. |
| 120 | 131 | 1101B | Raphanus sativus L. |
| 120 | 131 | 1016 | 1 Rorippa palustris (L.) Bess. ssp. glabra (O.E. Schulz) R. Stuckey  var. glabrata (Lunell) R. Stuckey |
| 120 | 131 | 1017 | 1 Rorippa palustris (L.) Bess. ssp. hispida (Desv.) Jonsell |
| 120 | 131 | 1015 | 1 Rorippa sessiliflora (Nutt.) A.S. Hitchc. |
| 120 | 131 | 1018 | 1 Rorippa sylvestris (L.) Bess. |
| 120 | 132 | 1038 | 1 Sibara virginica (L.) Rollins |
| 120 | 132 | 1007B | Sinapis alba L. |
| 120 | 132 | 1011 | 2 Sinapis arvensis L. |
| 120 | 132 | 1005 | 2 Sisymbrium altissimum L. |
| 120 | 132 | 1005A | Sisymbrium loeseli L. |
| 120 | 132 | 1004 | 2 Sisymbrium officinale (L.) Scop. var. leiocarpum DC. |
| 120 | 132 | 1003 | 2 Sisymbrium officinale (L.) Scop. var. officinale |
| 120 | 134 | 1001 | 2 Thlaspi arvense L. |
| 120 | 134 | 1002 | 2 Thlaspi perfoliatum L. |

## CACTACEAE

| | | | |
|---|---|---|---|
| 136 | 139 | 1458 | 1 Opuntia humifusa (Raf.) Raf. |

## CALLITRICHACEAE

| | | | |
|---|---|---|---|
| 141 | 141 | 1345 | 1 Callitriche heterophylla Pursh emend. Darby |
| 141 | 141 | 1344 | 1 Callitriche terrestris Raf. emend. Torr. |

## CAMPANULACEAE

| | | | |
|---|---|---|---|
| 141 | 142 | 1965 | 1 Campanula americana L. |
| 141 | 142 | 1966 | 1 Campanula aparinoides Pursh |
| 141 | 142 | 1965A | 2 Campanula rapunculoides L. |

## CAMPANULACEAE    CONT.

| | | | | |
|---|---|---|---|---|
| 141 | 142 | 1968 | 1 | Campanula rotundifolia L. |
| 141 | 144 | 1971 | 1 | Lobelia cardinalis L. ssp. cardinalis |
| 141 | 144 | 1975 | 1 | Lobelia inflata L. |
| 141 | 144 | 1974 | 1 | Lobelia kalmii L. |
| 141 | 144 | 1973 | 1 | Lobelia puberula Michx. |
| 141 | 144 | 1972 | 1 | Lobelia siphilitica L. |
| 141 | 144 | 1977B | 1 | Lobelia spicata Lam. var. campanulata McVaugh |
| 141 | 144 | 1977 | 1 | Lobelia spicata Lam. var. hirtella Gray |
| 141 | 144 | 1976 | 1 | Lobelia spicata Lam. var. leptostachya (A. DC.) Mackenzie & Bush |
| 141 | 144 | 1977A | 1 | Lobelia spicata Lam. var. spicata |
| 141 | 145 | 1970 | 1 | Triodanis perfoliata (L.) Nieuwl. |

## CAPPARIDACEAE

| | | | | |
|---|---|---|---|---|
| 146 | 147 | 1056 | 1 | Polanisia dodecandra (L.) DC. ssp. dodecandra |
| 146 | 147 | 1057 | 1 | Polanisia dodecandra (L.) DC. ssp. trachysperma (Torr. & Gray) Iltis |

## CAPRIFOLIACEAE

| | | | | |
|---|---|---|---|---|
| 147 | 147 | 1956 | 1 | Diervilla lonicera P. Mill. |
| 147 | 147 | 1950 | 1 | Linnaea borealis L. ssp. americana (Forbes) Hulten |
| 147 | 147 | 1951 | 1 | Lonicera canadensis Bartr. |
| 147 | 147 | 1953 | 1 | Lonicera dioica L. |
| 147 | 147 | 1954 | 1 | Lonicera dioica L. var. glaucescens (Rydb.) Butters |
| 147 | 147 | 1952 | 2 | Lonicera japonica Thunb. |
| 147 | 147 | 1955 | 1 | Lonicera prolifera (Kirchn.) Rehd. |
| 147 | 147 | 1955A4 | | Lonicera tatarica L. |
| 147 | 147 | 1955A | | Lonicera xylosteum L. |
| 147 | 147 | 1931 | 1 | Sambucus canadensis L. var. canadensis |
| 147 | 147 | 1932 | 1 | Sambucus racemosa L. ssp. pubens (Michx.) House |
| 147 | 148 | 0571A4 | | Symphoricarpos albus (L.) Blake var. laevigatus (Fern.) Blake |
| 147 | 148 | 1949C | | Symphoricarpos occidentalis Hook. |
| 147 | 148 | 1949 | 1 | Symphoricarpos orbiculatus Moench |
| 147 | 148 | 1948 | 1 | Triosteum angustifolium L. |
| 147 | 148 | 1945 | 1 | Triosteum aurantiacum Bickn. |
| 147 | 148 | 1947 | 1 | Triosteum aurantiacum Bickn. var. glaucescens Wieg. |
| 147 | 148 | 1946 | 1 | Triosteum aurantiacum Bickn. var. illinoense (Wieg.) Palmer & Steyermark |

CAPRIFOLIACEAE     CONT.

| | | | | |
|---|---|---|---|---|
| 147 | 148 | 1944 | 1 | Triosteum perfoliatum L. |
| 147 | 148 | 1934 | 1 | Viburnum acerifolium L. |
| 147 | 148 | 1935 | 1 | Viburnum cassinoides L. |
| 147 | 148 | 1942 | 1 | Viburnum dentatum L. var. deamii (Rehd.) Fern. |
| 147 | 148 | 1936A4 | | Viburnum lantana L. |
| 147 | 148 | 1936 | 1 | Viburnum lentago L. |
| 147 | 148 | 1941 | 1 | Viburnum molle Michx. |
| 147 | 148 | 1941A4 | | Viburnum opulus L. |
| 147 | 148 | 1937 | 1 | Viburnum prunifolium L. |
| 147 | 148 | 1939 | 1 | Viburnum rafinesquianum Schultes var. affine (Bush) House |
| 147 | 148 | 1940 | 1 | Viburnum rafinesquianum Schultes var. rafinesquianum |
| 147 | 148 | 1938 | 1 | Viburnum rufidulum Raf. |
| 147 | 148 | 1933 | 1 | Viburnum trilobum Marsh. |

CARYOPHYLLACEAE

| | | | | |
|---|---|---|---|---|
| 149 | 149 | 0913 | 2 | Agrostemma githago L. |
| 149 | 149 | 0906 | 2 | Arenaria serpyllifolia L. |
| 149 | 150 | 0902 | 2 | Cerastium arvense L. |
| 149 | 150 | 0900 | 2 | Cerastium fontanum Baumg. ssp. triviale (Link) Jalas |
| 149 | 150 | 0903 | 2 | Cerastium glomeratum Thuill. |
| 149 | 150 | 0904 | 2 | Cerastium nutans Raf. |
| 149 | 150 | 0904A | | Cerastium semidecandrum L. |
| 149 | 150 | 0925 | 2 | Dianthus armeria L. |
| 149 | 150 | 0925A | | Dianthus barbatus L. |
| 149 | 150 | 0924A | | Gypsophila acutifolia Stevens ex Spreng. |
| 149 | 150 | 0924B | | Gypsophila muralis L. |
| 149 | 150 | 0924A4 | | Gypsophila scorzonerifolia Ser. |
| 149 | 151 | 0904B | | Holosteum umbellatum L. |
| 149 | 151 | 0924C | | Lychnis chalcedonica L. |
| 149 | 151 | 0924D | | Lychnis coronaria (L.) Desr. |
| 149 | 151 | 0908A4 | | Minuartia michauxii (Fern.) Farw. var. michauxii |
| 149 | 151 | 0909 | 1 | Minuartia patula (Michx.) Mattf. |
| 149 | 151 | 0908 | 1 | Minuartia stricta (Sw.) Hiern |
| 149 | 152 | 0907 | 1 | Moehringia lateriflora (L.) Fenzl |
| 149 | 152 | 0910 | 1 | Paronychia canadensis (L.) Wood |
| 149 | 152 | 0911 | 1 | Paronychia fastigiata (Raf.) Fern. var. fastigiata |
| 149 | 153 | 0905 | 1 | Sagina decumbens (Ell.) Torr. & Gray |
| 149 | 153 | 0926 | 2 | Saponaria officinalis L. |
| 149 | 153 | 0912 | 2 | Scleranthus annuus L. |
| 149 | 153 | 0924 | 2 | Silene alba (P. Mill.) Krause |
| 149 | 153 | 0920 | 1 | Silene antirrhina L. |

## CARYOPHYLLACEAE CONT.

```
149  153  0920A   Silene armeria L.
149  153  0918   2 Silene cserei Baumg.
149  153  0919   2 Silene dichotoma Ehrh.
149  153  0916   1 Silene nivea (Nutt.) Otth
149  153  0921   2 Silene noctiflora L.
149  153  0922   1 Silene regia Sims
149  153  0914   1 Silene stellata (L.) Ait. f.
149  153  0923   1 Silene virginica L.
149  153  0917   2 Silene vulgaris (Moench) Garcke
149  155  0909A4  Spergula media (L.) C. Presl ex Griseb.
149  156  0909A   Spergularia rubra (L.) J. & C. Presl
149  156  0898   1 Stellaria corei Shinners
149  156  0895   2 Stellaria graminea L.
149  156  0896   1 Stellaria longifolia Muhl ex Willd.
149  156  0899   2 Stellaria media (L.) Vill.
149  156  0897   1 Stellaria pubera Michx.
149  157  0927   2 Vaccaria pyramidata Medic.
```

## CELASTRACEAE

```
157  157  1360A4  Celastrus orbiculatus Thunb.
157  157  1360   1 Celastrus scandens L.
157  157  1358   1 Euonymus americanus L.
157  157  1357   1 Euonymus atropurpureus Jacq.
157  157  1359   1 Euonymus obovatus Nutt.
```

## CERATOPHYLLACEAE

```
157  157  0933   1 Ceratophyllum demersum L.
```

## CHENOPODIACEAE

```
158  158  0875   1 Atriplex littoralis L.
158  158  0874   1 Atriplex patula L.
158  158  0875A  2 Atriplex rosea L.
158  159  0866   1 Chenopodium album L.
158  159  0860   2 Chenopodium ambrosioides L.
158  159  0864A4  Chenopodium berlandieri Moq. var.
                    boscianum (Moq.) H.A. Wahl
158  159  0864   1 Chenopodium berlandieri Moq. var.
                    zschackii (J. Murr) J. Murr
158  159  0862   2 Chenopodium botrys L.
158  159  0865   1 Chenopodium bushianum Aellen
158  159  0862B  1 Chenopodium capitatum (L.) Aschers.
158  159  0867   1 Chenopodium desiccatum A. Nels.
                    leptophylloides (J. Murr) H.A. Wahl
```

22

## CHENOPODIACEAE    CONT.

| | | | | |
|---|---|---|---|---|
| 158 | 159 | 0868 | 1 | Chenopodium gigantospermum Aellen |
| 158 | 159 | 0863 | 2 | Chenopodium glaucum L. var. glaucum |
| 158 | 159 | 0863A | | Chenopodium leptophyllum (Moq.) Nutt. ex S. Wats. |
| 158 | 159 | 0870 | 1 | Chenopodium missouriense Aellen |
| 158 | 159 | 0871 | 2 | Chenopodium murale L. |
| 158 | 159 | 0869 | 1 | Chenopodium standleyanum Aellen |
| 158 | 159 | 0872 | 2 | Chenopodium urbicum L. |
| 158 | 159 | 0867A | 2 | Chenopodium vulvaria L. |
| 158 | 160 | 0877 | 1 | Corispermum hyssopifolium L. |
| 158 | 160 | 0876 | 1 | Corispermum nitidum Kit. ex Schultes |
| 158 | 160 | 0873 | 1 | Cycloloma atriplicifolium (Spreng.) Coult. |
| 158 | 160 | 0875B | 2 | Kochia scoparia (L.) Schrad. |
| 158 | 161 | 0878 | 2 | Salsola kali L. |
| 158 | 161 | 0877A | | Suaeda depressa (Pursh) S. Wats. |

## CISTACEAE

| | | | | |
|---|---|---|---|---|
| 161 | 161 | 1424 | 1 | Helianthemum bicknellii Fern. |
| 161 | 161 | 1423 | 1 | Helianthemum canadense (L.) Michx. |
| 161 | 162 | 1425A | | Hudsonia tomentosa Nutt. |
| 161 | 162 | 1425 | 1 | Hudsonia tomentosa Nutt. var. intermedia Peck |
| 161 | 162 | 1427 | 1 | Lechea minor L. |
| 161 | 162 | 1431 | 1 | Lechea pulchella Raf. var. moniliformis (Bickn.) Seymour |
| 161 | 162 | 1428 | 1 | Lechea racemulosa Michx. |
| 161 | 162 | 1430 | 1 | Lechea stricta Leggett |
| 161 | 162 | 1429 | 1 | Lechea tenuifolia Michx. |
| 161 | 162 | 1426 | 1 | Lechea villosa Ell. |

## CLUSIACEAE

| | | | | |
|---|---|---|---|---|
| 162 | 162 | 1410 | 1 | Hypericum adpressum Bart. |
| 162 | 162 | 1414 | 1 | Hypericum boreale (Britt.) Bickn. |
| 162 | 162 | 1417 | 1 | Hypericum canadense L. |
| 162 | 162 | 1412 | 1 | Hypericum cistifolium Lam. |
| 162 | 162 | 1411 | 1 | Hypericum denticulatum Walt. var. denticulatum |
| 162 | 162 | 1409 | 1 | Hypericum dolabriforme Vent. |
| 162 | 162 | 1416 | 1 | Hypericum drummondii (Grev. & Hook.) Torr. & Gray |
| 162 | 162 | 1405 | 1 | Hypericum frondosum Michx. |
| 162 | 162 | 1415 | 1 | Hypericum gentianoides (L.) B.S.P. |
| 162 | 162 | 1404 | 1 | Hypericum kalmianum L. |
| 162 | 162 | 1413 | 1 | Hypericum majus (Gray) Britt. |

CLUSIACEAE        CONT.

| | | | | |
|---|---|---|---|---|
| 162 | 162 | 1418 | 1 | Hypericum mutilum L. |
| 162 | 162 | 1407 | 2 | Hypericum perforatum L. |
| 162 | 162 | 1406 | 1 | Hypericum prolificum L. |
| 162 | 162 | 1408 | 1 | Hypericum punctatum Lam. |
| 162 | 162 | 1403 | 1 | Hypericum pyramidatum Ait. |
| 162 | 162 | 1402 | 1 | Hypericum stragulum P. Adams & Robson |
| 162 | 163 | 1420 | 1 | Triandenum fraseri (Spach) Gleason |
| 162 | 163 | 1421 | 1 | Triandenum tubulosum (Walt.) Gleason |
| 162 | 162 | 1419 | 1 | Triandenum virginicum (L.) Raf. |
| 162 | 163 | 1422 | 1 | Triandenum walteri (J.G. Gmel.) Gleason |

COMMELINACEAE

| | | | | |
|---|---|---|---|---|
| 163 | 164 | 0589 | 1 | Commelina communis L. |
| 163 | 164 | 0590 | 1 | Commelina diffusa Burm. f. |
| 163 | 164 | 0592 | 1 | Commelina erecta L. var. deamiana Fern. |
| 163 | 164 | 3592A | 1 | Commelina erecta L. var. erecta |
| 163 | 164 | 0591 | 1 | Commelina virginica L. |
| 163 | 164 | 0593 | 1 | Tradescantia ohiensis Raf. var. ohiensis |
| 163 | 164 | 0594 | 1 | Tradescantia subaspera Ker-Gawl. var. subaspera |
| 163 | 164 | 0595 | 1 | Tradescantia virginiana L. |

CONVOLVULACEAE

| | | | | |
|---|---|---|---|---|
| 165 | 165 | 1672 | 1 | Calystegia fraterniflora (Mackenzie & Bush) Brummitt |
| 165 | 165 | 1670 | 2 | Calystegia pubescens Lindl. |
| 165 | 165 | 1673 | 1 | Calystegia sepium (L.) R. Br. ssp. americana (Sims) Brummitt |
| 165 | 165 | 1671 | 1 | Calystegia sepium (L.) R. Br. ssp. sepium |
| 165 | 165 | 1669 | 1 | Calystegia spithamaea (L.) Pursh ssp. spithamaea |
| 165 | 166 | 1674 | 2 | Convolvulus arvensis L. |
| 165 | 166 | 1664 | 1 | Cuscuta campestris Yuncker |
| 165 | 166 | 1666 | 1 | Cuscuta cephalanthii Engelm. |
| 165 | 166 | 1662 | 1 | Cuscuta compacta Juss. |
| 165 | 166 | 1667 | 1 | Cuscuta corylii Engelm. |
| 165 | 166 | 1660 | 1 | Cuscuta cuspidata Engelm. |
| 165 | 166 | 1661 | 1 | Cuscuta glomerata Choisy |
| 165 | 166 | 1665 | 1 | Cuscuta gronovii Willd. |
| 165 | 166 | 1665A | 1 | Cuscuta gronovii Willd. var. calyptrata Engelm. |
| 165 | 166 | 1663 | 1 | Cuscuta pentagona Engelm. |
| 165 | 166 | 1668 | 1 | Cuscuta polygonorum Engelm. |
| 165 | 167 | 1680 | 2 | Ipomoea coccinea L. |
| 165 | 167 | 1675 | 1 | Ipomoea lacunosa L. |

### CONVOLVULACEAE    CONT.

```
165  167  1678  2 Ipomoea nil (L.) Roth
165  167  1676  1 Ipomoea pandurata (L.) G.F.W. Mey.
```

### CORNACEAE

```
169  169  1556   1 Cornus alternifolia L. f.
169  169  1564   1 Cornus amomum P. Mill. ssp. amomum
169  169  1563   1 Cornus amomum P. Mill. ssp. obliqua
                    (Raf.) J.S. Wilson
169  169  1559   1 Cornus asperifolia Michx.
169  169  1554   1 Cornus canadensis L.
169  169  1554B    Cornus drummondii C.A. Mey.
169  169  1555   1 Cornus florida L.
169  169  1562   1 Cornus foemina P. Mill. ssp. foemina
169  169  1561   1 Cornus foemina P. Mill. ssp. racemosa
                    (Lam.) J.S. Wilson
169  169  1557   1 Cornus rugosa Lam.
169  169  1558   1 Cornus sericea L. ssp. sericea
```

### CRASSULACEAE

```
169  170  1061A    Sedum purpureum (L.) Schultes
169  170  1061B    Sedum sarmentosum Bunge
169  170  1062   1 Sedum telephioides Michx.
169  170  1063   1 Sedum ternatum Michx.
```

### CUCURBITACEAE

```
172  172  1961B    Citrullus colocynthis (L.) Schrad.
172  172  1962A    Cucurbita foetidissima H.B.K.
172  172  1963   1 Echinocystis lobata (Michx.) Torr. & Gray
172  173  1962   1 Melothria pendula L.
172  173  1964   1 Sicyos angulatus L.
```

### CYPERACEAE

```
173  173  0402   1 Bulbostylis capillaris (L.) C.B. Clarke
173  173  0499   1 Carex abscondita Mackenzie
173  173  0499A    Carex acutiformis Ehrh.
173  173  0432   1 Carex aggregata Mackenzie
173  173  0469   1 Carex alata Torr.
173  173  0468   1 Carex albolutescens Schwein.
173  173  0506   1 Carex albursina Sheldon
173  173  0443   1 Carex alopecoidea Tuckerman
173  173  0515   1 Carex amphibola Steud. var. amphibola
```

| | | | | |
|---|---|---|---|---|
| 173 | 173 | 0516A1 | | Carex amphibola Steud. var. rigida (Bailey) Fern. |
| 173 | 173 | 0516 | 1 | Carex amphibola Steud. var. turgida Fern. |
| 173 | 173 | 0434 | 1 | Carex annectens (Bickn.) Bickn. |
| 173 | 173 | 0538 | 1 | Carex aquatilis Wahlenb. var. aquatilis |
| 173 | 173 | 0434C | | Carex arctata Boott |
| 173 | 173 | 0477 | 1 | Carex artitecta Mackenzie var. artitecta |
| 173 | 173 | 0478 | 1 | Carex artitecta Mackenzie var. subtilirostris F.J. Herm. |
| 173 | 173 | 0552 | 1 | Carex atherodes Spreng. |
| 173 | 173 | 0452 | 1 | Carex atlantica Bailey var. incomperta (Bickn.) F.J. Herm. |
| 173 | 173 | 0492 | 1 | Carex aurea Nutt. |
| 173 | 173 | 0457 | 1 | Carex bebbii (Bailey) Fern. |
| 173 | 173 | 0463 | 1 | Carex bicknellii Britt. |
| 173 | 173 | 0507 | 1 | Carex blanda Dewey |
| 173 | 173 | 0462 | 1 | Carex brevior (Dewey) Mackenzie ex Lunell |
| 173 | 173 | 0455 | 2 | Carex bromoides Willd. |
| 173 | 173 | 0532 | 1 | Carex bushii Mackenzie |
| 173 | 173 | 0537 | 1 | Carex buxbaumii Wahlenb. |
| 173 | 173 | 0447 | 1 | Carex canescens L. ssp. arctiformis (Mackenzie) Calder & Taylor var. disjuncta Fern. |
| 173 | 173 | 0448 | 1 | Carex canescens L. ssp. canescens var. subloliacea (Laestad.) Hartman |
| 173 | 173 | 0497 | 1 | Carex careyana Dewey |
| 173 | 173 | 0531 | 1 | Carex caroliniana Schwein. |
| 173 | 173 | 0454 | 1 | Carex cephalantha (Bailey) Bickn. |
| 173 | 173 | 0431 | 1 | Carex cephaloidea (Dewey) Dewey |
| 173 | 173 | 0424 | 1 | Carex cephalophora Willd. |
| 173 | 173 | 0420 | 1 | Carex chordorrhiza Ehrh. ex L. f. |
| 173 | 173 | 0481 | 1 | Carex communis Bailey |
| 173 | 173 | 0547 | 1 | Carex comosa Boott |
| 173 | 173 | 0444 | 1 | Carex conjuncta Boott |
| 173 | 173 | 0514 | 1 | Carex conoidea Willd. |
| 173 | 173 | 0423 | 1 | Carex convoluta Mackenzie |
| 173 | 173 | 0511 | 1 | Carex crawei Dewey |
| 173 | 173 | 0544 | 1 | Carex crinita Lam. |
| 173 | 173 | 0471 | 1 | Carex cristatella Britt. |
| 173 | 173 | 0442 | 1 | Carex crus-corvi Schuttlw. ex Kunze |
| 173 | 173 | 0526 | 1 | Carex cryptolepis Mackenzie |
| 173 | 173 | 0466 | 1 | Carex cumulata (Bailey) Fern. |
| 173 | 173 | 0520 | 1 | Carex davisii Schwein. & Torr. |
| 173 | 173 | 0521 | 1 | Carex debilis Michx. var. debilis |
| 173 | 173 | 0522 | 1 | Carex debilis Michx. var. rudgei Bailey |
| 173 | 173 | 0436 | 1 | Carex decomposita Muhl. |
| 173 | 173 | 0437 | 1 | Carex diandra Schrank |
| 173 | 173 | 0500 | 1 | Carex digitalis Willd. |
| 173 | 173 | 0445 | 1 | Carex disperma Dewey |
| 173 | 173 | 0490 | 1 | Carex eburnea Boott |

CYPERACEAE        CONT.

| | | | | |
|---|---|---|---|---|
| 173 | 173 | 0479 | 1 | Carex emmonsii Dewey |
| 173 | 173 | 0540 | 1 | Carex emoryi Dewey |
| 173 | 173 | 0460 | 1 | Carex festucacea Willd. |
| 173 | 173 | 0517 | 1 | Carex flaccosperma Dewey var. glaucodea |

(Tuckerman) Kukenth.

| | | | | |
|---|---|---|---|---|
| 173 | 173 | 0527 | 1 | Carex flava L. |
| 173 | 173 | 0419 | 1 | Carex foenea Willd. var. foenea |
| 173 | 173 | 0545 | 1 | Carex folliculata L. |
| 173 | 173 | 0554 | 1 | Carex frankii Kunth |
| 173 | 173 | 0491 | 1 | Carex garberi Fern. |
| 173 | 173 | 0569 | 1 | Carex gigantea Rudge |
| 173 | 173 | 0508 | 1 | Carex gracilescens Steud. |
| 173 | 173 | 0518 | 1 | Carex gracillima Schwein. |
| 173 | 173 | 0510 | 1 | Carex granularis Muhl. ex Willd. var. |

granularis

173  173  0509  1 Carex granularis Muhl. ex Willd. var.

haleana (Olney) Porter

| | | | | |
|---|---|---|---|---|
| 173 | 173 | 0429 | 1 | Carex gravida Bailey var. gravida |
| 173 | 173 | 0430 | 1 | Carex gravida Bailey var. lunelliana |

(Mackenzie) F.J. Herm.

| | | | | |
|---|---|---|---|---|
| 173 | 173 | 0563 | 1 | Carex grayi Carey |
| 173 | 173 | 0539 | 1 | Carex haydenii Dewey |
| 173 | 173 | 0482 | 1 | Carex heliophila Mackenzie |
| 173 | 173 | 0530 | 1 | Carex hirsutella Mackenzie |
| 173 | 173 | 0489 | 1 | Carex hirtifolia Mackenzie |
| 173 | 173 | 0513 | 1 | Carex hitchcockiana Dewey |
| 173 | 173 | 0451 | 1 | Carex howei Mackenzie |
| 173 | 173 | 0550 | 1 | Carex hyalinolepis Steud. |
| 173 | 173 | 0546 | 1 | Carex hystricina Muhl. ex Willd. |
| 173 | 173 | 0450 | 1 | Carex interior Bailey |
| 173 | 173 | 0564 | 1 | Carex intumescens Rudge |
| 173 | 173 | 0476 | 1 | Carex jamesii Schwein. |
| 173 | 173 | 0549 | 1 | Carex lacustris Willd. |
| 173 | 173 | 0441 | 1 | Carex laevivaginata (Kukenth.) Mackenzie |
| 173 | 173 | 0533 | 1 | Carex lanuginosa Michx. |
| 173 | 173 | 0534 | 1 | Carex lasiocarpa Ehrh. |
| 173 | 173 | 0534A | | Carex lasiocarpa Ehrh. var. americana |

Fern.

173  173  0502  1 Carex laxiculmis Schwein. var. copulata

(Bailey) Fern.

| | | | | |
|---|---|---|---|---|
| 173 | 173 | 0501 | 1 | Carex laxiculmis Schwein. var. laxiculmis |
| 173 | 173 | 0504 | 1 | Carex laxiflora Lam. var. laxiflora |
| 173 | 173 | 0505 | 1 | Carex laxiflora Lam. var. serrulata F.J. |

Herm.

| | | | | |
|---|---|---|---|---|
| 173 | 173 | 0425 | 1 | Carex leavenworthii Dewey |
| 173 | 173 | 0474 | 1 | Carex leptalea Wahlenb. ssp. harperi |

(Fern.) Calder & Taylor

| | | | | |
|---|---|---|---|---|
| 173 | 173 | 0473 | 1 | Carex leptalea Wahlenb. var. leptalea |
| 173 | 173 | 0473A | | Carex leptonervia Fern. |

27

```
173  173  0536  1 Carex limosa L.
173  173  0467  1 Carex longii Mackenzie
173  173  0566  1 Carex louisianica Bailey
173  173  0568  1 Carex lupuliformis Sartwell ex Dewey
173  173  0567  1 Carex lupulina Willd.
173  173  0562  1 Carex lurida Wahlenb.
173  173  0494  1 Carex meadii Dewey
173  173  0426  1 Carex mesochorea Mackenzie
173  173  0461  1 Carex molesta Mackenzie
173  173  0428  1 Carex muhlenbergii Willd. var. enervis
                   Boott
173  173  0427  1 Carex muhlenbergii Willd. var.
                   muhlenbergii
173  173  0472  1 Carex muskingumensis Schwein.
173  173  0480  1 Carex nigromarginata Schwein.
173  173  0459  1 Carex normalis Mackenzie
173  173  0512  1 Carex oligocarpa Willd.
173  173  0561  1 Carex oligosperma Michx.
173  173  0561A   Carex pallescens L.
173  173  0561B1  Carex pedunculata Willd.
173  173  0483  1 Carex pensylvanica Lam.
173  173  0488  1 Carex picta Steud.
173  173  0496  1 Carex plantaginea Lam.
173  173  0498  1 Carex platyphylla Carey
173  173  0438  1 Carex prairea Dewey
173  173  0519  1 Carex prasina Wahlenb.
173  173  0519A4  Carex projecta Mackenzie
173  173  0548  1 Carex pseudocyperus L.
173  173  0421  1 Carex retroflexa Willd.
173  173  0421A   Carex retroflexa Willd. var. texensis
                   (Torr.) Fern.
173  173  0560  1 Carex retrorsa Schwein.
173  173  0487  1 Carex richardsonii R. Br.
173  173  0422  1 Carex rosea Willd.
173  173  0558  1 Carex rostrata Stokes ex With.
173  173  0485  1 Carex rugosperma Mackenzie
173  173  0418  1 Carex sartwellii Dewey
173  173  0418A 1 Carex sartwellii Dewey var.
                   stenorrhyncha F.J. Herm.
173  173  0418B   Carex scabrata Schwein.
173  173  0456  1 Carex scoparia Schkuhr ex Willd.
173  173  0449  1 Carex seorsa  Howe
173  173  0535  1 Carex shortiana Dewey
173  173  0535A5  Carex socialis Mohlenbrock & Schwegm.
173  173  0433  1 Carex sparganioides Willd.
173  173  0523  1 Carex sprengelii Dewey ex Spreng.
173  173  0556  1 Carex squarrosa L.
173  173  0453  1 Carex sterilis Willd.
173  173  0440  1 Carex stipata Muhl. ex Willd. var.
                   maxima Chapman
```

CYPERACEAE     CONT.

```
173  173  0439  1 Carex stipata Muhl. ex Willd. var.
                   stipata
173  173  0465  1 Carex straminea Willd.
173  173  0541  1 Carex stricta Lam. var. stricta
173  173  0542  1 Carex stricta Lam. var. strictior
                   (Dewey) Carey
173  173  0503  1 Carex styloflexa Buckl.
173  173  0464  1 Carex suberecta (Olney) Britt.
173  173  0528  1 Carex swanii (Fern.) Mackenzie
173  173  0458  1 Carex tenera Dewey
173  173  0493  1 Carex tetanica Schkuhr
173  173  0486  1 Carex tonsa (Fern.) Bickn.
173  173  0543  1 Carex torta Boott
173  173  0470  1 Carex tribuloides Wahlenb.
173  173  0553  1 Carex trichocarpa Schkuhr
173  173  0446  1 Carex trisperma Dewey
173  173  0559  1 Carex tuckermanii Dewey
173  173  0555  1 Carex typhina Michx.
173  173  0484  1 Carex umbellata Schkuhr ex Willd.
173  173  0557  1 Carex vesicaria L.
173  173  3529  1 Carex virescens Willd.
173  173  0524  1 Carex viridula Michx.
173  173  0525  1 Carex viridula Michx. f. intermedia
                   (Dudley) F.J. Herm.
173  173  0435  1 Carex vulpinoidea Michx.
173  173  0435A1  Carex vulpinoidea Michx. var.
                   pycnocephala F.J. Herm.
173  173  0475  1 Carex willdenowii Schkuhr
173  173  0495  1 Carex woodii Dewey
173  173  0551  1 Carex X subimpressa Clokey.
173  181  0405  1 Cladium mariscoides (Muhl.) Torr.
173  181  0345  1 Cyperus acuminatus Torr. & Hook.
173  181  0344  1 Cyperus aristatus Rottb. var. aristatus
173  181  0350  1 Cyperus dentatus Torr.
173  181  0342  1 Cyperus diandrus Torr.
173  181  0353  1 Cyperus engelmannii Steud.
173  181  0355  1 Cyperus erythrorhizos Muhl.
173  181  0357  1 Cyperus esculentus L.
173  181  0348  1 Cyperus filiculmis Vahl var. filiculmis
173  181  0349  1 Cyperus filiculmis Vahl var. macilentus
                   Fern.
173  181  0341  1 Cyperus flavescens L.
173  181  0351  1 Cyperus houghtonii Torr.
173  181  0356  1 Cyperus odoratus L. var. odoratus
173  181  0347  1 Cyperus ovularis (Michx.) Torr.
173  181  0346  1 Cyperus pseudovegetus Steud.
173  181  0343  1 Cyperus rivularis Kunth
173  181  0352  1 Cyperus schweinitzii Torr.
173  181  0354A 1 Cyperus strigosus L. var. multiflorus
                   Geise.
```

CYPERACEAE          CONT.

| 173 | 181 | 0358 | 1 Cyperus tenuifolius (Steud.) Dandy |
| 173 | 183 | 0340 | 1 Dulichium arundinaceum (L.) Britt. |
| 173 | 183 | 0391 | 1 Eleocharis acicularis (L.) Roemer & Schultes var. acicularis |
| 173 | 183 | 0395 | 1 Eleocharis elliptica Kunth var. elliptica |
| 173 | 183 | 0388 | 1 Eleocharis engelmannii Steud. |
| 173 | 183 | 0380 | 1 Eleocharis equisetoides (Ell.) Torr. |
| 173 | 183 | 0390 | 1 Eleocharis erythropoda Steud. |
| 173 | 183 | 0384 | 1 Eleocharis geniculata (L.) Roemer & Schultes |
| 173 | 183 | 0386 | 1 Eleocharis intermedia Schultes |
| 173 | 183 | 0386A | Eleocharis macrostachya Britt. |
| 173 | 183 | 0393 | 1 Eleocharis melanocarpa Torr. |
| 173 | 183 | 0394 | 1 Eleocharis microcarpa Torr. |
| 173 | 183 | 0387 | 1 Eleocharis obtusa (Willd.) Schultes |
| 173 | 183 | 0385 | 1 Eleocharis obtusa (Willd.) Schultes var. ovata (Roth) Drapalik & Mohlenbrock |
| 173 | 183 | 0383 | 1 Eleocharis olivacea Torr. |
| 173 | 183 | 0399 | 1 Eleocharis pauciflora (Lightf.) Link |
| 173 | 183 | 0381 | 1 Eleocharis quadrangulata (Michx.) Roemer & Schultes |
| 173 | 183 | 0382 | 1 Eleocharis robbinsii Oakes |
| 173 | 183 | 0398 | 1 Eleocharis rostellata (Torr.) Torr. |
| 173 | 183 | 0389 | 1 Eleocharis smallii Britt. |
| 173 | 183 | 0396 | 1 Eleocharis verrucosa (Svens.) L. Harms |
| 173 | 183 | 0392 | 1 Eleocharis wolfii Gray |
| 173 | 184 | 0361 | 1 Eriophorum angustifolium Honckeny |
| 173 | 184 | 0360 | 1 Eriophorum gracile W.D.J. Koch |
| 173 | 184 | 0359 | 1 Eriophorum vaginatum L. ssp. spissum (Fern.) Hulten |
| 173 | 184 | 0363 | 1 Eriophorum virginicum L. |
| 173 | 184 | 0362 | 1 Eriophorum viridicarinatum (Engelm.) Fern. |
| 173 | 185 | 0401 | 1 Fimbristylis autumnalis (L.) Roemer & Schultes |
| 173 | 185 | 0400 | 1 Fimbristylis caroliniana (Lam.) Fern. |
| 173 | 185 | 0400A | Fimbristylis puberula (Michx.) Vahl var. puberula |
| 173 | 185 | 0364 | 1 Fuirena pumila (Torr.) Spreng. |
| 173 | 185 | 0339 | 1 Hemicarpha drummondii Nees |
| 173 | 185 | 0338 | 1 Hemicarpha micrantha (Vahl) Britt. |
| 173 | 186 | 0403 | 1 Psilocarya nitens (Vahl) Wood |
| 173 | 186 | 0404 | 1 Psilocarya scirpoides Torr. |
| 173 | 186 | 0406 | 1 Rhynchospora alba (L.) Vahl |
| 173 | 186 | 0407 | 1 Rhynchospora capillacea Torr. |
| 173 | 186 | 0411 | 1 Rhynchospora capitellata (Michx.) Vahl |
| 173 | 186 | 0409 | 1 Rhynchospora corniculata (Lam.) Gray var. interior Fern. |
| 173 | 186 | 0410 | 1 Rhynchospora globularis (Chapm.) Small var. globularis |

## CYPERACEAE          CONT.

```
173 186 0412   1 Rhynchospora macrostachya Gray
173 187 0372   1 Scirpus acutus Muhl. ex Bigelow
173 187 0369   1 Scirpus americanus Pers.
173 187 0374   1 Scirpus atrovirens Willd.
173 187 0378   1 Scirpus cyperinus (L.) Kunth
173 187 0373   1 Scirpus fluviatilis (Torr.) Gray
173 187 0375   1 Scirpus georgianus Harper
173 187 0375A5   Scirpus hallii Gray
173 187 0377   1 Scirpus lineatus Michx.
173 187 0379   1 Scirpus pedicellatus Fern.
173 187 0376   1 Scirpus polyphyllus Vahl
173 187 0366   1 Scirpus purshianus Fern.
173 187 0367   1 Scirpus smithii Gray
173 187 0365   1 Scirpus subterminalis Torr.
173 187 0371   1 Scirpus tabernaemontanii K.C. Gmel.
173 187 0370   1 Scirpus torreyi Olney
173 188 0413   1 Scleria oligantha Michx.
173 188 0413A  1 Scleria pauciflora Muhl. ex Willd. var.
                   caroliniana (Willd.) Wood
173 188 0414A4   Scleria reticularis Michx. var. pubscens
                   Britt.
173 188 0414   1 Scleria reticularis Michx. var.
                   reticularis
173 188 0416   1 Scleria triglomerata Michx.
173 188 0417   1 Scleria verticillata Muhl. ex Willd.
```

## DIOSCOREACEAE

```
189 189 0677A3   Dioscorea batatas Dcne.
189 189 0677   1 Dioscorea hirticaulis Bartlett
189 189 0679   1 Dioscorea quaternata (Walt.) J.F. Gmel.
189 189 0678   1 Dioscorea villosa L.
```

## DIPSACACEAE

```
189 189 1961   2 Dipsacus fullonum L.
189 189 1961A    Dipsacus laciniatus L.
```

## DROSERACEAE

```
189 189 1060   1 Drosera intermedia Hayne
189 189 1059   1 Drosera rotundifolia L.
```

EBENACEAE

190 190 1607 1 Diospyros virginiana L.

ELAEAGNACEAE

190 190 1460 1 Shepherdia canadensis (L.) Nutt.

ERICACEAE

```
191 191 1575 1 Andromeda polifolia L. var. glaucophylla
                 (Link.) DC.
191 191 1580 1 Arctostaphylos uva-ursi (L.) Spreng.
                 ssp. coactilis (Fern. & Macbr.) Love, Love & Kapoo
191 192 .1576 1 Chamaedaphne calyculata (L.) Moench
191 192 1565 1 Chimaphila maculata (L.) Pursh
191 192 1566 1 Chimaphila umbellata (L.) Bart. ssp.
                 cisatlantica (Blake) Hulten
191 193 1578 1 Epigaea repens L.
191 193 1579 1 Gaultheria procumbens L.
191 193 1581 1 Gaylussacia baccata (Wang.) K. Koch
191 193 1574 1 Kalmia latifolia L.
191 194 1573 1 Monotropa hypopithys L.
191 194 1572 1 Monotropa uniflora L.
191 194 1567 1 Orthilia secunda (L.) House ssp. secunda
191 194 1577 1 Oxydendrum arboreum (L.) DC.
191 195 1570 1 Pyrola americana Sweet
191 195 1571 1 Pyrola asarifolia Michx. var. purpurea
                 (Bunge) Fern.
191 195 1569 1 Pyrola chlorantha Sw. var. chlorantha
191 195 1568 1 Pyrola elliptica Nutt.
191 195 1586 1 Vaccinium angustifolium Ait.
191 195 1584 1 Vaccinium arboreum Marsh.
191 195 1585 1 Vaccinium corymbosum L.
191 195 1589 1 Vaccinium macrocarpon Ait.
191 195 1588 1 Vaccinium myrtilloides Michx.
191 195 1590 1 Vaccinium oxycoccos L.
191 195 1587 1 Vaccinium pallidum Ait.
191 195 1582 1 Vaccinium stamineum L.
```

ERIOCAULACEAE

197 197 0588 1 Eriocaulon septangulare With.

EUPHORBIACEAE

197 197 1324 1 Acalypha deamii (Weatherby) Ahles

EUPHORBIACEAE     CONT.

| | | | | |
|---|---|---|---|---|
| 197 | 197 | 1326 | 1 | Acalypha gracilens Gray |
| 197 | 197 | 1322 | 2 | Acalypha ostryifolia Riddell |
| 197 | 197 | 1323 | 1 | Acalypha rhomboidea Raf. |
| 197 | 197 | 1325 | 1 | Acalypha virginica L. |
| 197 | 198 | 1327A | | Chamaesyce geyeri (Engelm.) Small |
| 197 | 198 | 1331 | 1 | Chamaesyce glyptosperma (Engelm.) Small |
| 197 | 198 | 1332 | 1 | Chamaesyce humistrata (Engelm. ex Gray) Small |
| 197 | 198 | 1335 | 1 | Chamaesyce maculata (L.) Small |
| 197 | 198 | 1329 | 1 | Chamaesyce polygonifolia (L.) Small |
| 197 | 198 | 1330 | 1 | Chamaesyce serpens (H.B.K.) Small |
| 197 | 198 | 1334 | 1 | Chamaesyce vermiculata (Raf.) House |
| 197 | 201 | 1319 | 2 | Croton capitatus Michx. |
| 197 | 201 | 1318 | 2 | Croton glandulosus L. var. septentrionalis Muell.-Arg. |
| 197 | 201 | 1320 | 2 | Croton monanthogynus Michx. |
| 197 | 201 | 1321 | 1 | Crotonopsis elliptica Willd. |
| 197 | 202 | 1343 | 1 | Euphorbia commutata Engelm. |
| 197 | 202 | 1336 | 1 | Euphorbia corollata L. |
| 197 | 202 | 1341 | 2 | Euphorbia cyparissias L. |
| 197 | 202 | 1340 | 2 | Euphorbia esula L. |
| 197 | 202 | 1328 | 2 | Euphorbia marginata Pursh |
| 197 | 202 | 1339 | 1 | Euphorbia obtusata Pursh |
| 197 | 202 | 1342 | 2 | Euphorbia peplus L. |
| 197 | 203 | 1317 | 1 | Phyllanthus caroliniensis Walt. |
| 197 | 204 | 1337 | 1 | Poinsettia dentata (Michx.) Klotzsch & Garcke |
| 197 | 204 | 1338 | 1 | Poinsettia heterophylla (L.) Klotzsch & Garcke |
| 197 | 204 | 1327 | 1 | Tragia cordata Michx. |

FABACEAE

| | | | | |
|---|---|---|---|---|
| 205 | 205 | 1225 | 1 | Amorpha canescens Pursh |
| 205 | 205 | 1226 | 1 | Amorpha fruticosa L. |
| 205 | 206 | 1277 | 1 | Amphicarpaea bracteata (L.) Fern. |
| 205 | 206 | 1276A4 | | Apios americana Medic. |
| 205 | 206 | 1233 | 1 | Astragalus canadensis L. |
| 205 | 212 | 1208A4 | | Baptisia australis (L.) R. Br. |
| 205 | 212 | 1209 | 1 | Baptisia lactea (Raf.) Thieret |
| 205 | 212 | 1208B5 | | Baptisia leucophaea Nutt. |
| 205 | 212 | 1208 | 1 | Baptisia tinctoria (L.) R. Br. var. crebra Fern. |
| 205 | 214 | 1196 | 1 | Cassia fasciculata Michx. |
| 205 | 214 | 1199 | 1 | Cassia hebecarpa Fern. |
| 205 | 214 | 1200 | 1 | Cassia marilandica L. |
| 205 | 214 | 1194 | 1 | Cassia nictitans L. |
| 205 | 214 | 1198 | 1 | Cassia occidentalis L. |

33

```
205   215   1193   1 Cercis canadensis L.
205   215   1205   1 Cladrastis kentukea (Dum.-Cours.) Rudd
205   215   1276   1 Clitoria mariana L.
205   215   1234A4   Coronilla varia L.
205   215   1210   2 Crotalaria sagittalis L.
205   216   1228   1 Dalea candida (Michx.) Willd.
205   216   1226C  2 Dalea leporina (Ait.) Bullock
205   216   1227   1 Dalea purpurea Vent.
205   217   1192   2 Desmanthus illinoensis (Michx.) MacM. ex
                     B.L. Robins. & Fern.
205   217   1245   1 Desmodium canadense (L.) DC.
205   217   1241   1 Desmodium canescens (L.) DC.
205   217   1251   1 Desmodium ciliare (Muhl. ex Willd.) DC.
205   217   1243   1 Desmodium cuspidatum (Muhl. ex Willd.)
                     Loud. var. cuspidatum
205   217   1244   1 Desmodium cuspidatum (Muhl. ex Willd.)
                     Loud. var. longifolium (Torr. & Gray) Schub.
205   217   1240   1 Desmodium glutinosum (Muhl. ex Willd.)
                     Wood
205   217   1242   1 Desmodium illinoense Gray
205   217   1247   1 Desmodium laevigatum (Nutt.) DC.
205   217   1250   1 Desmodium marilandicum (L.) DC.
205   217   1238   1 Desmodium nudiflorum (L.) DC.
205   217   1252   1 Desmodium obtusum (Muhl. ex Willd.) DC.
205   217   1246   1 Desmodium paniculatum (L.) DC.
205   217   1239   1 Desmodium pauciflorum (Nutt.) DC.
205   217   1248   1 Desmodium perplexum Schub.
205   217   1236   1 Desmodium rotundifolium DC.
205   217   1237   1 Desmodium sessilifolium (Torr.) Torr. &
                     Gray
205   217   1249   1 Desmodium viridiflorum (L.) DC.
205   219   1201   1 Gleditsia aquatica Marsh.
205   219   1202   1 Gleditsia triacanthos L.
205   219   1203   1 Gleditsia X texana Sarg.
205   219   1233A    Glycyrrhiza lepidota (Nutt.) Pursh
205   220   1204   1 Gymnocladus dioicus (L.) K. Koch
205   220   1254   2 Kummerowia stipulacia (Maxim.) Makino
205   220   1253   2 Kummerowia striata (Thunb.) Schindl.
205   220   1270   1 Lathyrus japonicus Willd. var. glaber
                     (Ser.) Fern.
205   220   1270A    Lathyrus latifolius L.
205   220   1269   1 Lathyrus ochroleucus Hook.
205   220   1275   1 Lathyrus palustris L. var. myrtifolius
                     (Muhl.) Gray
205   220   1274   1 Lathyrus palustris L. var. palustris
205   220   1271   1 Lathyrus venosus Muhl. ex Willd.
205   220   1272   1 Lathyrus venosus Muhl. ex Willd. ssp.
                     venosus var. intonsus (Butters & St. John)
                     C.L. Hitchc.
205   221   1255C1   Lespedeza bicolor Turcz.
```

FABACEAE        CONT.

| | | | | |
|---|---|---|---|---|
| 205 | 221 | 1255 | 1 | Lespedeza capitata Michx. |
| 205 | 221 | 1255A | | Lespedeza cuneata (Dum.-Cours.) G. Don |
| 205 | 221 | 1257 | 1 | Lespedeza hirta (L.) Hornem. |
| 205 | 221 | 1260 | 1 | Lespedeza intermedia (S. Wats.) Britt. |
| 205 | 221 | 1261 | 1 | Lespedeza intermedia (S. Wats) f. hahnii (Blake) Hopkins |
| 205 | 221 | 1265 | 1 | Lespedeza procumbens Michx. |
| 205 | 221 | 1262 | 1 | Lespedeza repens (L.) Bart. |
| 205 | 221 | 1264 | 1 | Lespedeza stuevei Nutt. |
| 205 | 221 | 1263 | 1 | Lespedeza violacea (L.) Pers. |
| 205 | 221 | 1258 | 1 | Lespedeza virginica (L.) Britt. |
| 205 | 221 | 1259 | 1 | Lespedeza virginica (L.) Britt. f. deamii Hopkins |
| 205 | 221 | 1265A | 1 | Lespedeza X brittonii Bickn. |
| 205 | 221 | 1255B | 1 | Lespedeza X longifolia DC. |
| 205 | 221 | 1256 | 1 | Lespedeza X nuttallii Darl. |
| 205 | 222 | 1221A | | Lotus corniculatus L. |
| 205 | 222 | 1221B | | Lotus purshianus (Benth.) Clem. & Clem. |
| 205 | 223 | 1211 | 2 | Lupinus perennis L. |
| 205 | 230 | 1213 | 2 | Medicago lupulina L. |
| 205 | 230 | 1212 | 2 | Medicago sativa L. |
| 205 | 230 | 1214 | 2 | Melilotus alba Medic. |
| 205 | 230 | 1215 | 2 | Melilotus officinalis (L.) Pallas |
| 205 | 232 | 1281 | 1 | Phaseolus polystachyus (L.) B.S.P. |
| 205 | 233 | 1223 | 1 | Psoralea onobrychis Nutt. |
| 205 | 233 | 1224 | 1 | Psoralea psoralioides (Walt.) Cory var. eglandulosa (Ell.) Freeman |
| 205 | 233 | 1222A | 1 | Psoralea stipulata Torr. & Gray |
| 205 | 233 | 1222 | 2 | Psoralea tenuiflora Pursh |
| 205 | 234 | 1232A | | Robinia hispida L. |
| 205 | 234 | 1232 | 1 | Robinia pseudoacacia L. |
| 205 | 234 | 1232B | | Robinia viscosa Vent. ex Vauq. |
| 205 | 235 | 1282 | 1 | Strophostyles helvola (L.) Ell. |
| 205 | 235 | 1284 | 1 | Strophostyles leiosperma (Torr. & Gray) Piper |
| 205 | 235 | 1283 | 1 | Strophostyles umbellata (Muhl. ex Willd.) Britt. |
| 205 | 235 | 1235 | 1 | Stylosanthes biflora (L.) B.S.P. |
| 205 | 235 | 1229 | 1 | Tephrosia virginiana (L.) Pers. |
| 205 | 236 | 1221A4 | | Trifolium arvense L. |
| 205 | 236 | 1221 | 1 | Trifolium aureum Pollich. |
| 205 | 236 | 1220A | 2 | Trifolium dubium Sibthorp |
| 205 | 236 | 1221B4 | | Trifolium hybridum L. |
| 205 | 236 | 1221C4 | | Trifolium pratense L. |
| 205 | 236 | 1220 | 2 | Trifolium procumbens L. |
| 205 | 236 | 1219 | 1 | Trifolium reflexum L. |
| 205 | 236 | 1221D4 | | Trifolium repens L. |
| 205 | 238 | 1268 | 1 | Vicia americana Muhl. ex Willd. |
| 205 | 238 | 1267 | 1 | Vicia caroliniana Walt. |

## FABACEAE CONT.

```
205  238  1267A4   Vicia cracca L.
205  238  1266    2 Vicia villosa Roth
205  238  1231    1 Wisteria macrostachya Nutt.
```

## FAGACEAE

```
239  239  0779   1 Castanea dentata (Marsh.) Borkh.
239  239  0778   1 Fagus grandifolia Ehrh.
239  239  0780   1 Quercus alba L.
239  239  0782   1 Quercus bicolor Willd.
239  239  0795   1 Quercus coccinea Muenchh.
239  239  0795A  1 Quercus coccinea Muenchh. var.
                    tuberculata Sarg.
239  239  0794   1 Quercus ellipsoidalis E. J. Hill
239  239  0796   1 Quercus falcata Michx.
239  239  0789   1 Quercus imbricaria Michx.
239  239  0788   1 Quercus lyrata Walt.
239  239  0787   1 Quercus macrocarpa Michx.
239  239  0797   1 Quercus marilandica Muenchh.
239  239  0784   1 Quercus montana Willd.
239  239  0783   1 Quercus muhlenbergii Engelm.
239  239  0796B  1 Quercus pagoda Raf.
239  239  0792   1 Quercus palustris Muenchh.
239  239  0781   1 Quercus prinoides Willd.
239  239  0790   1 Quercus rubra L. var. rubra
239  239  0793   1 Quercus shumardii Buckl.
239  239  0793A  1 Quercus shumardii Buckl. var. schneckii
                    (Britt.) Sarg.
239  239  0786   1 Quercus stellata Wang.
239  239  0791   1 Quercus velutina Lam.
239  239  0780A  1 Quercus X beadlei Trel.
239  239  0797A  1 Quercus X bushii Sarg.
239  239  0780B  1 Quercus X deamii Trel.
239  239  0789A  1 Quercus X exacta Trel.
239  239  0780C  1 Quercus X fernowii Trel.
239  239  0780D  1 Quercus X jackiana Schneid.
239  239  0789B  1 Quercus X leana Nutt.
239  239  0782A  1 Quercus X schuettei Trel.
```

## GENTIANACEAE

```
242  242  1620   1 Bartonia virginica (L.) B.S.P.
242  242  1619B    Centaurium erythraea Rafn.
242  242  1619A    Centaurium pulchellum (Sw.) Druce
242  243  1630   1 Frasera caroliniensis Walt.
242  243  1628   1 Gentiana alba Muhl.
242  243  1625   1 Gentiana andrewsii Griseb.
```

## GENTIANACEAE     CONT.

```
242  243  1627  1  Gentiana puberulenta J. Pringle
242  243  1626  1  Gentiana saponaria L.
242  243  1629  1  Gentiana villosa L.
242  244  1624  1  Gentianella quinquefolia (L.) Small ssp.
                      occidentalis (Gray) J. Gillett
242  244  1622  1  Gentianopsis crinita (Froel.) Ma
242  244  1623  1  Gentianopsis procera (Holm) Ma
242  245  1621  1  Obolaria virginica L.
242  245  1618  1  Sabatia angularis (L.) Pursh
242  245  1619  1  Sabatia campanulata (L.) Torr.
```

## GERANIACEAE

```
245  245  1291A    Erodium cicutarium (L.) L'Her.
245  246  1288  1  Geranium bicknellii Britt. var.
                      bicknellii
245  246  1289  1  Geranium carolinianum L. var.
                      carolinianum
245  246  1290  1  Geranium carolinianum L. var.
                      confertiflorum Fern.
245  246  1287  2  Geranium columbinum L.
245  246  1285  1  Geranium maculatum L.
245  246  1291  2  Geranium pusillum L.
245  246  1286  1  Geranium robertianum L.
```

## HALORAGIDACEAE

```
248  249  1500  1  Myriophyllum exalbescens Fern.
248  249  1501  1  Myriophyllum heterophyllum Michx.
248  249  1502  1  Myriophyllum pinnatum (Walt.) B.S.P.
248  249  1503  1  Myriophyllum verticillatum L.
248  249  1504  1  Proserpinaca palustris L. var.
                      amblyogona Fern.
248  249  1505  1  Proserpinaca palustris L. var. crebra
                      Fern. & Grisc.
```

## HAMAMELIDACEAE

```
249  249  1084  1  Hamamelis virginiana L.
249  249  1083  1  Liquidambar styraciflua L.
```

## HIPPOCASTANACEAE

```
249  249  1370  1  Aesculus flava Soland.
249  249  1369  1  Aesculus glabra Willd.
```

## HIPPOCASTANACEAE CONT.

249 249 1368B 2 Aesculus hippocastanum L.

## HIPPURIDACEAE

249 249 1506 1 Hippuris vulgaris L.

## HYDROCHARITACEAE

249 250 0116 1 Elodea canadensis L.C. Rich.
249 250 0117 1 Elodea nuttallii (Planch.) St. John
249 250 0118 1 Vallisneria americana Michx.

## HYDROPHYLLACEAE

250 250 1696 1 Ellisia nyctelea L.
250 250 1692 1 Hydrophyllum appendiculatum Michx.
250 250 1693 1 Hydrophyllum canadense L.
250 250 1695 1 Hydrophyllum macrophyllum Nutt.
250 250 1694 1 Hydrophyllum virginianum L.
250 251 1697 1 Phacelia bipinnatifida Michx.
250 251 1699 1 Phacelia purshii Buckl.
250 251 1698 1 Phacelia ranunculacea (Nutt.) Constance

## IRIDACEAE

254 254 0684 2 Belamcanda chinensis (L.) DC.
254 254 0682 1 Iris brevicaulis Raf.
254 254 0681 1 Iris cristata Soland.
254 254 0681A   Iris pseudacorus L.
254 254 0683 1 Iris virginica L. var. shrevei (Small)
                 E. Anders.
254 256 0685 1 Sisyrinchium albidum Raf.
254 256 0686 1 Sisyrinchium angustifolium P. Mill.
254 256 0687 1 Sisyrinchium atlanticum Bickn.
254 256 0686A   Sisyrinchium montanum Greene

## JUGLANDACEAE

256 256 0759 1 Carya cordiformis (Wang.) K. Koch
256 256 0763 1 Carya glabra (P. Mill.) Sweet
256 256 0763A 1 Carya glabra (P. Mill.) Sweet var.
                 megacarpa (Sarg.) Sarg.
256 256 0758 1 Carya illinoensis (Wang.) K. Koch
256 256 0761 1 Carya laciniosa (Michx. f.) Loud.

JUGLANDACEAE     CONT.

| 256 | 256 | 0764 | 1 | Carya ovalis (Wang.) Sarg. |
| 256 | 256 | 0760 | 1 | Carya ovata (P. Mill.) K. Koch |
| 256 | 256 | 0765 | 1 | Carya pallida (Ashe) Engl. & Graebn. |
| 256 | 256 | 0766 | 1 | Carya texana Buckl. |
| 256 | 256 | 0762 | 1 | Carya tomentosa (Poir.) Nutt. |
| 256 | 257 | 0756 | 1 | Juglans cinerea L. |
| 256 | 257 | 0757 | 1 | Juglans nigra L. |

JUNCACEAE

| 257 | 257 | 0615 | 1 | Juncus acuminatus Michx. |
| 257 | 257 | 0621A | 1 | Juncus alpinus Vill. ssp. alpinus |
| 257 | 257 | 0621 | 1 | Juncus alpinus Vill. ssp. nodulosus (Wahlenb.) Lindm. var. rariflorus Hartman |
| 257 | 257 | 0620 | 1 | Juncus articulatus L. |
| 257 | 257 | 0600 | 1 | Juncus balticus Willd. var. littoralis Engelm. |
| 257 | 257 | 0600A | 1 | Juncus balticus Willd. var. littoralis Engelm. f. dissitiflorus Engelm. |
| 257 | 257 | 0614 | 1 | Juncus brachycarpus Engelm. |
| 257 | 257 | 0611 | 1 | Juncus brachycephalus (Engelm.) Buch. |
| 257 | 257 | 0601 | 1 | Juncus bufonius L. |
| 257 | 257 | 0610 | 2 | Juncus canadensis J. Gay ex Laharpe |
| 257 | 257 | 0612 | 1 | Juncus diffusissimus Buckl. |
| 257 | 257 | 0599A | 1 | Juncus effusus L. var. pylaei (Laharpe) Fern. & Wieg. |
| 257 | 257 | 0599 | 1 | Juncus effusus L. var. solutus Fern. & Wieg. |
| 257 | 257 | 0602 | 1 | Juncus gerardii Loisel. |
| 257 | 257 | 0603 | 1 | Juncus greenei Oakes & Tuckerman |
| 257 | 257 | 0606 | 1 | Juncus interior Wieg. |
| 257 | 257 | 0608 | 1 | Juncus marginatus Rostk. |
| 257 | 257 | 0608A | | Juncus militaris Bigelow |
| 257 | 257 | 0616 | 1 | Juncus nodatus Coville |
| 257 | 257 | 0618 | 1 | Juncus nodosus L. |
| 257 | 257 | 0617 | 1 | Juncus pelocarpus E. Mey. |
| 257 | 257 | 0613 | 1 | Juncus scirpoides Lam. |
| 257 | 257 | 0604 | 1 | Juncus secundus Beauv. ex Poir. |
| 257 | 257 | 0605 | 1 | Juncus tenuis Willd. var. tenuis |
| 257 | 257 | 0607 | 1 | Juncus tenuis Willd. var. uniflorus (Farw.) Farw. |
| 257 | 257 | 0605A | 1 | Juncus tenuis Willd. var. williamsii Fern. |
| 257 | 257 | 0619 | 1 | Juncus torreyi Coville |
| 257 | 259 | 0622 | 1 | Luzula acuminata Raf. var. acuminata |
| 257 | 259 | 0624 | 1 | Luzula bulbosa (Wood) Rydb. |
| 257 | 259 | 0625 | 1 | Luzula echinata (Small) F.J. Herm. var. echinata |

257  259  0625A 1 Luzula echinata (Small) F.J. Herm. var.
                   mesochorea F.J. Herm.
257  259  0623  1 Luzula multiflora (Retz.) Lej.


LAMIACEAE

260  260  1742  1 Agastache nepetoides (L.) Kuntze
260  260  1743  1 Agastache scrophulariifolia (Willd.)
                   Kuntze
260  261  1726A   Ajuga reptans L.
260  261  1769  1 Blephilia ciliata (L.) Benth.
260  261  1770  1 Blephilia hirsuta (Pursh) Benth.
260  261  1775  1 Calamintha arkansana (Nutt.) Shinners
260  261  1774  1 Clinopodium vulgare L.
260  261  1793  1 Collinsonia canadensis L.
260  261  1780  1 Cunila origanoides (L.) Britt.
260  261  1747A4   Dracocephalum parviflorum Nutt.
260  262  1745  2 Glechoma hederacea L. var. hederacea
260  262  1746  2 Glechoma hederacea L. var. micrantha
                   Moric.
260  262  1772  1 Hedeoma hispidum Pursh
260  262  1771  1 Hedeoma pulegioides (L.) Pers.
260  262  1751  2 Lamium amplexicaule L.
260  262  1752  1 Lamium purpureum L.
260  262  1753  2 Leonurus cardiaca L.
260  262  1753A   Leonurus marrubiastrum L.
260  263  1785  1 Lycopus americanus Muhl. ex Bart.
260  263  1783  1 Lycopus amplectens Raf.
260  263  1786B   Lycopus asper Greene
260  263  1784  1 Lycopus rubellus Moench
260  263  1781  1 Lycopus uniflorus Michx.
260  263  1782  1 Lycopus virginicus L.
260  263  1741  2 Marrubium vulgare L.
260  263  1773  2 Melissa officinalis L.
260  263  1791  1 Mentha arvensis L.
260  263  1792B   Mentha arvensis L. ssp. haplocalyx Briq.
260  263  1792  1 Mentha canadensis L.
260  263  1787  2 Mentha spicata L.
260  263  1792A 2 Mentha X gentilis L.
260  263  1788  2 Mentha X piperita L.
260  263  1790  2 Mentha X rotundifolia (L.) Huds.
260  263  1764  1 Monarda bradburiana Beck
260  263  1765  1 Monarda clinopodia L.
260  263  1766  1 Monarda fistulosa L. var. fistulosa
260  263  1767  1 Monarda fistulosa L. var. mollis (L.)
                   Benth.
260  263  1768  1 Monarda punctata L. var. villicaulis

| | | | |
|---|---|---|---|
| 260 | 264 | 1744 | 2 Nepeta cataria L. |
| 260 | 264 | 1794 | 2 Perilla frutescens (L.) Britt. var. crispa (Benth.) Deane |
| 260 | 265 | 1748A1 | Physostegia parviflora Nutt. ex Gray |
| 260 | 265 | 1748 | 1 Physostegia virginiana (L.) Benth. |
| 260 | 265 | 1747A | 2 Prunella vulgaris L. |
| 260 | 265 | 1747 | 1 Prunella vulgaris L. var. elongata Benth. |
| 260 | 266 | 1777 | 1 Pycnanthemum flexuosum (Walt.) B.S.P. |
| 260 | 266 | 1777A1 | Pycnanthemum incanum (L.) Michx. |
| 260 | 266 | 1778 | 1 Pycnanthemum pilosum Nutt. |
| 260 | 266 | 1776 | 1 Pycnanthemum pycnanthemoides (Leavenworth) Fern. |
| 260 | 266 | 1776A1 | Pycnanthemum torrei Benth. |
| 260 | 266 | 1779 | 1 Pycnanthemum virginianum (L.) Durand & Jackson |
| 260 | 266 | 1761A | Salvia azurea Michx. & Lam. var. grandiflora Benth. |
| 260 | 266 | 1761 | 1 Salvia lyrata L. |
| 260 | 266 | 1763 | 2 Salvia nemorosa L. |
| 260 | 266 | 1762 | 2 Salvia reflexa Hornem. |
| 260 | 267 | 1774A | 2 Satureja hortensis L. |
| 260 | 267 | 1735 | 1 Scutellaria australis (Fassett) Epling |
| 260 | 267 | 1739 | 1 Scutellaria elliptica Muhl. var. elliptica |
| 260 | 267 | 1739A | Scutellaria elliptica Muhl. var. hirsuta (Short & Peter) Fern. |
| 260 | 267 | 1732 | 1 Scutellaria galericulata L. |
| 260 | 267 | 1738 | 1 Scutellaria incana Biehler |
| 260 | 267 | 1736 | 1 Scutellaria lateriflora L. |
| 260 | 267 | 1735A | 1 Scutellaria leonardii Epling |
| 260 | 267 | 1733 | 1 Scutellaria nervosa Pursh |
| 260 | 267 | 1740 | 1 Scutellaria ovata Hill |
| 260 | 267 | 1734 | 1 Scutellaria parvula Michx. |
| 260 | 267 | 1737 | 1 Scutellaria saxatilis Riddell |
| 260 | 268 | 1755 | 1 Stachys aspera Michx. |
| 260 | 268 | 1760 | 1 Stachys clingmanii Small |
| 260 | 268 | 1754 | 1 Stachys hyssopifolia Michx. |
| 260 | 268 | 1759 | 1 Stachys nuttallii Shuttlw. ex Benth. |
| 260 | 268 | 1757 | 1 Stachys palustris L. var. homotricha Fern. |
| 260 | 268 | 1756 | 1 Stachys tenuifolia Willd. var. hispida (Pursh.) Fern. |
| 260 | 268 | 1758 | 1 Stachys tenuifolia Willd. var. tenuifolia |
| 260 | 269 | 1750 | 1 Synandra hispidula (Michx.) Baill. |
| 260 | 269 | 1726 | 1 Teucrium canadense L. |
| 260 | 269 | 1728 | 1 Teucrium canadense L. var. boreale (Bickn.) Shinners |
| 260 | 269 | 1727 | 1 Teucrium canadense L. var. virginicum (L.) Eat. |

LAMIACEAE          CONT.

260  269  1779A4  Thymus praecox Opiz
260  269  1779A   Thymus praecox Opiz ssp. arcticus (Dur.)
                  Jalas
260  269  1730   1 Trichostema brachiatum L.
260  269  1731   1 Trichostema dichotomum L.


LARDIZABALACEAE

269  269  0981D3  Akebia quinata (Houtt.) Dcne.


LAURACEAE

269  270  0989   1 Lindera benzoin (L.) Blume
269  270  0988   1 Sassafras albidum (Nutt.) Nees


LEMNACEAE

270  270  0582A   Lemna gibba L.
270  270  0580   1 Lemna minor L.
270  270  0579   1 Lemna minuta H.B.K.
270  270  0581   1 Lemna perpusilla Torr.
270  270  0581A4  Lemna trisulca L.
270  270  0582   1 Lemna valdiviana Phil.
270  270  0577   1 Spirodela polyrhiza (L.) Schleid.
270  270  0584   1 Wolffia borealis (Engelm.) Landolt
270  270  0583A  1 Wolffia braziliensis Weddell
270  270  0583   1 Wolffia columbiana Karst.
270  270  0584A4  Wolffia punctata Griseb.
270  270  0585   1 Wolffiella gladiata (Hegelm.) Hegelm.


LENTIBULARIACEAE

271  271  1885   1 Utricularia cornuta Michx.
271  271  1886   1 Utricularia gibba L.
271  271  1888   1 Utricularia intermedia Hayne
271  271  1890   1 Utricularia macrorhiza Le Conte
271  271  1889   1 Utricularia minor L.
271  271  1883   1 Utricularia purpurea Walt.
271  271  1891   1 Utricularia radiata Small
271  271  1884   1 Utricularia resupinata B.D. Greene
271  271  1884A   Utricularia vulgaris L.


42

## LILIACEAE

| | | | | |
|---|---|---|---|---|
| 271 | 271 | 0665 | 1 | Aletris farinosa L. |
| 271 | 271 | 0638C4 | | Allium burdickii (Hanes) A.G. Jones |
| 271 | 271 | 0638A4 | | Allium canadense L. var. canadense |
| 271 | 271 | 0640 | 1 | Allium cernuum Roth |
| 271 | 271 | 0638 | 2 | Allium sativum L. |
| 271 | 271 | 0638A | | Allium schoenoprasum L. |
| 271 | 271 | 0638B4 | | Allium stellatum Ker-Gawl. |
| 271 | 271 | 0636 | 1 | Allium tricoccum Ait. |
| 271 | 271 | 0637 | 2 | Allium vineale L. |
| 271 | 273 | 0650 | 2 | Asparagus officinalis L. |
| 271 | 274 | 0648 | 1 | Camassia scilloides (Raf.) Cory |
| 271 | 274 | 0627 | 1 | Chamaelirium luteum (L.) Gray |
| 271 | 274 | 0651 | 1 | Clintonia borealis (Ait.) Raf. |
| 271 | 274 | 0657B | | Convallaria majalis L. |
| 271 | 275 | 0646 | 1 | Erythronium albidum Nutt. |
| 271 | 275 | 0647 | 1 | Erythronium americanum Ker-Gawl. |
| 271 | 276 | 0635 | 1 | Hemerocallis fulva (L.) L. |
| 271 | 276 | 0635A | | Hemerocallis lilio-asphodelus L. emend. Hyl. |
| 271 | 276 | 0674 | 1 | Hymenocallis caroliniana (L.) Herbert |
| 271 | 276 | 0676 | 1 | Hypoxis hirsuta (L.) Coville |
| 271 | 276 | 0644 | 1 | Lilium canadense L. var. canadense |
| 271 | 276 | 0645A | 2 | Lilium kelloggii Purdy |
| 271 | 276 | 0645B | | Lilium lancifolium Thunb. |
| 271 | 276 | 0645 | 1 | Lilium michiganense Farw. |
| 271 | 276 | 0642 | 1 | Lilium philadelphicum L. var. andinum (Nutt.) Ker-Gawl. |
| 271 | 276 | 0643 | 1 | Lilium superbum L. |
| 271 | 277 | 0655A4 | | Maianthemum canadense Desf. var. canadense |
| 271 | 277 | 0655 | 1 | Maianthemum canadense Desf. var. interius Fern. |
| 271 | 277 | 0658 | 1 | Medeola virginiana L. |
| 271 | 277 | 0631 | 1 | Melanthium virginicum L. |
| 271 | 277 | 0649E | | Muscari atlanticum Boiss & Reut. |
| 271 | 277 | 0649A | 2 | Muscari botryoides (L.) P. Mill. |
| 271 | 277 | 0641 | 1 | Nothoscordum bivalve (L.) Britt. |
| 271 | 277 | 0649 | 1 | Ornithogalum umbellatum L. |
| 271 | 277 | 0657 | 1 | Polygonatum biflorum (Walt.) Ell. |
| 271 | 277 | 0656 | 1 | Polygonatum pubescens (Willd.) Pursh |
| 271 | 278 | 0652 | 1 | Smilacina racemosa (L.) Desf. |
| 271 | 278 | 0652A4 | | Smilacina stellata (L.) Desf. |
| 271 | 278 | 0628 | 1 | Stenanthium gramineum (Ker-Gawl.) Morong |
| 271 | 278 | 0629 | 1 | Stenanthium gramineum (Ker-Gawl.) Morong var. robustum (S. Wats.) Fern. |
| 271 | 278 | 0626 | 1 | Tofieldia glutinosa (Michx.) Pers. |
| 271 | 278 | 0663 | 1 | Trillium cernuum L. var. macranthum Eames & Wieg. |
| 271 | 278 | 0663A | | Trillium flexipes Raf. |

## LILIACEAE          CONT.

| | | | | |
|---|---|---|---|---|
| 271 | 278 | 0662 | 1 | Trillium grandiflorum (Michx.) Salisb. |
| 271 | 278 | 0661 | 1 | Trillium nivale Riddell |
| 271 | 278 | 0660 | 1 | Trillium recurvatum Beck |
| 271 | 278 | 0659 | 1 | Trillium sessile L. |
| 271 | 279 | 0633 | 1 | Uvularia grandiflora Sm. |
| 271 | 279 | 0634 | 1 | Uvularia sessilifolia L. |
| 271 | 280 | 0632 | 1 | Veratrum woodii J.W. Robbins |
| 271 | 280 | 0630 | 1 | Zigadenus elegans Pursh ssp. glaucus (Nutt.) Hulten |

## LIMNANTHACEAE

| | | | | |
|---|---|---|---|---|
| 280 | 280 | 1346 | 1 | Floerkea proserpinacoides Willd. |

## LINACEAE

| | | | | |
|---|---|---|---|---|
| 280 | 281 | 1301 | 1 | Linum intercursum Bickn. |
| 280 | 281 | 1304 | 1 | Linum medium (Planch.) Britt. var. texanum (Planch.) Fern. |
| 280 | 281 | 1303 | 1 | Linum striatum Walt. |
| 280 | 281 | 1300 | 1 | Linum sulcatum Riddell |
| 280 | 281 | 1300A | | Linum usitatissimum L. |
| 280 | 281 | 1302 | 1 | Linum virginianum L. |

## LOGANIACEAE

| | | | | |
|---|---|---|---|---|
| 283 | 284 | 1617 | 1 | Spigelia marilandica L. |

## LORANTHACEAE

| | | | | |
|---|---|---|---|---|
| 284 | 284 | 0819 | 1 | Phoradendron serotinum (Raf.) M.C. Johnston |

## LYTHRACEAE

| | | | | |
|---|---|---|---|---|
| 285 | 285 | 1463 | 1 | Ammannia coccinea Rottb. |
| 285 | 285 | 1467 | 1 | Cuphea viscosissima Jacq. |
| 285 | 285 | 1468 | 1 | Decodon verticillatus (L.) Ell. |
| 285 | 285 | 1469 | 1 | Decodon verticillatus (L.) Ell. var. laevigatus Torr. & Gray |
| 285 | 285 | 1464 | 1 | Didiplis diandra (DC.) Wood |
| 285 | 285 | 1465 | 1 | Lythrum alatum Pursh |
| 285 | 285 | 1466 | 2 | Lythrum salicaria L. |
| 285 | 285 | 1461 | 1 | Rotala ramosior (L.) Koehne |

## MAGNOLIACEAE

| | | | | |
|---|---|---|---|---|
| 285 | 285 | 0986 | 1 | Liriodendron tulipifera L. |
| 285 | 285 | 0985 | 1 | Magnolia acuminata (L.) L. |
| 285 | 285 | 0985A1 | | Magnolia tripetala L. |

## MALVACEAE

| | | | | |
|---|---|---|---|---|
| 286 | 286 | 1389 | 2 | Abutilon theophrasti Medic. |
| 286 | 287 | 1389A | | Alcea rosea L. |
| 286 | 287 | 1394A1 | | Callirhoe involucrata (Nutt. ex Torr. & Gray) Gray |
| 286 | 287 | 1394 | 1 | Callirhoe triangulata (Leavenworth) Gray |
| 286 | 287 | 1397 | 1 | Hibiscus laevis All. |
| 286 | 287 | 1400 | 1 | Hibiscus lasiocarpus Cav. |
| 286 | 287 | 1398 | 1 | Hibiscus moscheutos L. |
| 286 | 287 | 1401 | 2 | Hibiscus trionum L. |
| 286 | 289 | 1393 | 2 | Malva moschata L. |
| 286 | 289 | 1392 | 2 | Malva neglecta Wallr. |
| 286 | 289 | 1391 | 2 | Malva rotundifolia L. |
| 286 | 289 | 1390 | 2 | Malva sylvestris L. |
| 286 | 289 | 1395 | 1 | Napaea dioica L. |
| 286 | 289 | 1396 | 1 | Sida spinosa L. |

## MARTYNIACEAE

| | | | | |
|---|---|---|---|---|
| 291 | 292 | 1877A | 1 | Proboscidea louisianica (P. Mill.) Thellung |

## MELASTOMATACEAE

| | | | | |
|---|---|---|---|---|
| 292 | 292 | 1471 | 1 | Rhexia mariana L. var. mariana |
| 292 | 292 | 1470 | 1 | Rhexia virginica L. |

## MENISPERMACEAE

| | | | | |
|---|---|---|---|---|
| 293 | 293 | 0984 | 1 | Calycocarpum lyonii (Pursh) Gray |
| 293 | 293 | 0983 | 1 | Cocculus carolinus (L.) DC. |
| 293 | 293 | 0982 | 1 | Menispermum canadense L. |

## MENYANTHACEAE

| | | | | |
|---|---|---|---|---|
| 293 | 293 | 1631 | 1 | Menyanthes trifoliata L. var. minor Raf. |

## MORACEAE

| | | | | |
|---|---|---|---|---|
| 293 | 294 | 0811 | 2 | Cannabis sativa L. ssp. sativa var. sativa |
| 293 | 294 | 0809 | 1 | Humulus japonicus Sieb. & Zucc. |
| 293 | 294 | 0810 | 1 | Humulus lupulus L. var. lupuloides E. Small |
| 293 | 294 | 0808 | 2 | Maclura pomifera (Raf.) Schneid. |
| 293 | 294 | 0807 | 2 | Morus alba L. |
| 293 | 294 | 0806 | 1 | Morus rubra L. |

## MYRICACEAE

| | | | | |
|---|---|---|---|---|
| 294 | 294 | 0755 | 1 | Comptonia peregrina (L.) Coult. |

## NAJADACEAE

| | | | | |
|---|---|---|---|---|
| 297 | 297 | 0099 | 1 | Najas flexilis (Willd.) Rostk. & Schmidt. |
| 297 | 297 | 0101 | 1 | Najas gracillima (A. Braun) Magnus |
| 297 | 297 | 0100 | 1 | Najas guadalupensis (Spreng.) Magnus |
| 297 | 297 | 0100A5 | | Najas marina L. |
| 297 | 297 | 0100A1 | | Najas minor All. |

## NYCTAGINACEAE

| | | | | |
|---|---|---|---|---|
| 297 | 298 | 0888A | | Mirabilis hirsuta (Pursh) MacM. |
| 297 | 298 | 0888B | | Mirabilis jalapa L. |
| 297 | 298 | 0888C | | Mirabilis linearis (Pursh) Heimerl |
| 297 | 298 | 0889 | 1 | Mirabilis nyctaginea (Michx.) MacM. |

## NYMPHAEACEAE

| | | | | |
|---|---|---|---|---|
| 299 | 299 | 0929 | 1 | Brasenia schreberi J.F. Gmel. |
| 299 | 299 | 0928A | 1 | Cabomba caroliniana Gray |
| 299 | 299 | 0928 | 1 | Nelumbo lutea (Willd.) Pers. |
| 299 | 299 | 0931 | 1 | Nuphar luteum (L.) Sibthorp & Sm. ssp. variegatum (Dur.) E.O. Beal |
| 299 | 300 | 0930 | 1 | Nymphaea odorata Ait. var. odorata |

## NYSSACEAE

| | | | | |
|---|---|---|---|---|
| 300 | 300 | 1552 | 1 | Nyssa sylvatica Marsh. var. sylvatica |

## OLEACEAE

| | | | | |
|---|---|---|---|---|
| 300 | 300 | 1616 | 1 | Forestiera acuminata (Michx.)  Poir. |
| 300 | 301 | 1609 | 1 | Fraxinus americana L. |
| 300 | 301 | 1610 | 1 | Fraxinus americana L. var. biltmoreana (Beadle) J. Wright ex Fern. |
| 300 | 301 | 1615 | 1 | Fraxinus nigra Marsh. |
| 300 | 301 | 1611 | 1 | Fraxinus pennsylvanica Marsh. |
| 300 | 301 | 1613 | 1 | Fraxinus profunda (Bush) Bush |
| 300 | 301 | 1614 | 1 | Fraxinus quadrangulata Michx. |
| 300 | 301 | 1616A4 | | Ligustrum obtusifolium Sieb. & Zucc. |
| 300 | 302 | 1616B | | Syringa vulgaris L. |

## ONAGRACEAE

| | | | | |
|---|---|---|---|---|
| 302 | 302 | 1487A | | Calylophus serrulatus (Nutt.) Raven |
| 302 | 304 | 1499 | 1 | Circaea alpina L. |
| 302 | 304 | 1498 | 1 | Circaea lutetiana (L.) Aschers. & Magnus ssp. canadensis (L.) Aschers. & Magnus |
| 302 | 305 | 1479 | 1 | Epilobium angustifolium L. ssp. angustifolium |
| 302 | 305 | 1483 | 1 | Epilobium ciliatum Raf. ssp. ciliatum |
| 302 | 305 | 1482 | 1 | Epilobium coloratum Biehler |
| 302 | 305 | 1483A | | Epilobium hirsutum L. |
| 302 | 305 | 1483B | | Epilobium leptophyllum Raf. |
| 302 | 305 | 1480 | 1 | Epilobium strictum Muhl. |
| 302 | 306 | 1495 | 1 | Gaura biennis L. |
| 302 | 306 | 1496 | 2 | Gaura coccinea Pursh |
| 302 | 306 | 1497 | 1 | Gaura filipes Spach |
| 302 | 306 | 1494 | 2 | Gaura parviflora Dougl. |
| 302 | 307 | 1475 | 1 | Ludwigia alternifolia L. |
| 302 | 307 | 1472 | 1 | Ludwigia decurrens Walt. |
| 302 | 307 | 1476 | 1 | Ludwigia glandulosa Walt. |
| 302 | 307 | 1474 | 1 | Ludwigia palustris (L.) Ell. |
| 302 | 307 | 1473 | 1 | Ludwigia peploides (H.B.K.) Raven ssp. glabrescens (Kuntze) Raven |
| 302 | 307 | 1477 | 1 | Ludwigia polycarpa Short & Peter |
| 302 | 307 | 1478 | 1 | Ludwigia sphaerocarpa Ell. |
| 302 | 308 | 1484 | 1 | Oenothera biennis L. ssp. biennis |
| 302 | 308 | 1490 | 1 | Oenothera fruticosa L. ssp. fruticosa |
| 302 | 308 | 1488 | 1 | Oenothera laciniata Hill |
| 302 | 308 | 1486A | 1 | Oenothera parviflora L. var. parviflora |
| 302 | 308 | 1491 | 1 | Oenothera perennis L. |
| 302 | 308 | 1489 | 1 | Oenothera pilosella Raf. |
| 302 | 308 | 1487 | 1 | Oenothera rhombipetala Nutt. ex Torr. & Gray |
| 302 | 308 | 1492 | 2 | Oenothera speciosa Nutt. |
| 302 | 308 | 1486 | 1 | Oenothera villosa Thunb. ssp. canovirens (Steele) D. Dietr. & Raven |

## ORCHIDACEAE

| | | | | |
|---|---|---|---|---|
| 310 | 310 | 0727 | 1 | Aplectrum hyemale (Muhl. ex Willd.) Nutt. |
| 310 | 310 | 0710 | 1 | Arethusa bulbosa L. |
| 310 | 310 | 0718 | 1 | Calopogon tuberosus (L.) B.S.P. |
| 310 | 310 | 0694 | 1 | Coeloglossum viride (L.) Hartman |
| 310 | 310 | 0694A4 | | Coeloglossum viride (L.) Hartman var. virescens (Muhl. ex Willd.) Luer |
| 310 | 311 | 0720 | 1 | Corallorhiza maculata (Raf.) Raf. |
| 310 | 311 | 0721 | 1 | Corallorhiza odontorhiza (Willd.) Nutt. |
| 310 | 311 | 0720A | | Corallorhiza trifida (L.) Chatelain |
| 310 | 311 | 0719 | 1 | Corallorhiza wisteriana Conrad |
| 310 | 311 | 0692 | 1 | Cypripedium acaule Ait. |
| 310 | 311 | 0689 | 1 | Cypripedium candidum Muhl. ex Willd. |
| 310 | 311 | 0690 | 1 | Cypripedium parviflorum Salisb. |
| 310 | 311 | 0691 | 1 | Cypripedium pubescens Willd. |
| 310 | 311 | 0688 | 1 | Cypripedium reginae Walt. |
| 310 | 312 | 0711 | 2 | Epipactis helleborine (L.) Crantz |
| 310 | 312 | 0693 | 1 | Galearis spectabilis (L.) Raf. |
| 310 | 312 | 0717 | 1 | Goodyera pubescens (Willd.) R. Br. |
| 310 | 312 | 0726 | 1 | Hexalectris spicata (Walt.) Barnh. |
| 310 | 313 | 0709 | 1 | Isotria verticillata (Muhl. ex Willd.) Raf. |
| 310 | 313 | 0723 | 1 | Liparis lilifolia (L.) L. C. Rich. ex Lindl. |
| 310 | 313 | 0724 | 1 | Liparis loeselii (L.) L. C. Rich. |
| 310 | 313 | 0722 | 1 | Malaxis unifolia Michx. |
| 310 | 314 | 0702 | 1 | Platanthera ciliaris (L.) Lindl. |
| 310 | 314 | 0699 | 1 | Platanthera clavellata (Michx.) Luer |
| 310 | 314 | 0697 | 1 | Platanthera dilatata (Pursh) Lindl. ex Beck |
| 310 | 314 | 0695 | 1 | Platanthera flava (L.) Lindl. var. flava |
| 310 | 314 | 0693A | | Platanthera flava (L.) Lindl. var. herbiola (R. Br.) Luer |
| 310 | 314 | 0701 | 1 | Platanthera hookeri (Torr. ex Gray) Lindl. |
| 310 | 314 | 0698 | 1 | Platanthera hyperborea (L.) Lindl. |
| 310 | 314 | 0703 | 1 | Platanthera lacera (Michx.) G. Don |
| 310 | 314 | 0704 | 1 | Platanthera leucophaea (Nutt.) Lindl. |
| 310 | 314 | 0700 | 1 | Platanthera orbiculata (Pursh) Lindl. |
| 310 | 314 | 0706 | 1 | Platanthera peramoena (Gray) Gray |
| 310 | 314 | 0705 | 1 | Platanthera psycodes (L.) Lindl. |
| 310 | 315 | 0707 | 1 | Pogonia ophioglossoides (L.) Juss. |
| 310 | 315 | 0716 | 1 | Spiranthes cernua (L.) L.C. Rich. |
| 310 | 315 | 0712 | 1 | Spiranthes lacera (Raf.) Raf. var. gracilis (Bigelow) Luer |

ORCHIDACEAE      CONT.

| 310 | 315 | 0714  | 1 | Spiranthes lucida (H.H. Eat.) Ames |
| 310 | 315 | 0714A4 |  | Spiranthes magnicamporum Sheviak |
| 310 | 315 | 0714A3 |  | Spiranthes ochroleuca (Rydb.) Rydb. |
| 310 | 315 | 0715  | 1 | Spiranthes ovalis Lindl. |
| 310 | 316 | 0725  | 1 | Tipularia discolor (Pursh) Nutt. |
| 310 | 316 | 0708  | 1 | Triphora trianthophora (Sw.) Rydb. |

OROBANCHACEAE

| 316 | 317 | 1878 | 1 | Conopholis americana (L.) Wallr. |
| 316 | 317 | 1882 | 1 | Epifagus virginiana (L.) Bart. |
| 316 | 317 | 1881 | 1 | Orobanche fasciculata Nutt. var. fasciculata |
| 316 | 317 | 1879 | 1 | Orobanche ludoviciana Nutt. ssp. ludoviciana |
| 316 | 317 | 1880 | 1 | Orobanche uniflora L. var. uniflora |

OXALIDACEAE

| 317 | 317 | 1294A | 2 | Oxalis corniculata L. |
| 317 | 317 | 1296  | 1 | Oxalis dillenii Jacq. ssp. filipes (Small) Eiten |
| 317 | 317 | 1297  | 1 | Oxalis europaea Jord. |
| 317 | 317 | 1298  | 1 | Oxalis europaea Jord. f. cymosa (Small) Wieg. |
| 317 | 317 | 1299  | 1 | Oxalis europaea Jord. f. villicaulis Wieg. |
| 317 | 317 | 1294  | 1 | Oxalis grandis Small |
| 317 | 317 | 1295A | 1 | Oxalis stricta L. var. piletocarpa Wieg. |
| 317 | 317 | 1295  | 1 | Oxalis stricta L. var. stricta |
| 317 | 317 | 1293  | 1 | Oxalis violacea L. var. trichophora Fassett |
| 317 | 317 | 1292  | 1 | Oxalis violacea L. var. violacea |

PAPAVERACEAE

| 318 | 318 | 0994A | 1 | Adlumia fungosa (Ait.) Greene ex B.S.P. |
| 318 | 318 | 0992A |  | Argemone polyanthemos (Fedde) G.B. Ownbey |
| 318 | 319 | 0992  | 2 | Chelidonium majus L. |
| 318 | 319 | 0996  | 1 | Corydalis flavula (Raf.) DC. |
| 318 | 319 | 0995  | 1 | Corydalis sempervirens (L.) Pers. |
| 318 | 319 | 0993  | 1 | Dicentra canadensis (Goldie) Walp. |
| 318 | 319 | 0994  | 1 | Dicentra cucullaria (L.) Bernh. |
| 318 | 320 | 0996A |  | Fumaria officinalis L. |
| 318 | 321 | 0990  | 1 | Sanguinaria canadensis L. |
| 318 | 321 | 0991  | 1 | Stylophorum diphyllum (Michx.) Nutt. |

## PASSIFLORACEAE

321 321 1457 1 Passiflora incarnata L.
321 321 1456 1 Passiflora lutea L. var. glabriflora
             Fern.

## PHYTOLACCACEAE

321 321 0890 1 Phytolacca americana L.

## PLANTAGINACEAE

323 323 1901A   Plantago arenaria Waldst. & Kit.
323 323 1901  1 Plantago aristata Michx.
323 323 1898  1 Plantago cordata Lam.
323 323 1902  2 Plantago lanceolata L.
323 323 1902A 2 Plantago lanceolata L. var.
               sphaerostachya Mert. & Koch
323 323 1899  1 Plantago major L.
323 323 1903  1 Plantago patagonica Jacq. var. patagonica
323 323 1905  1 Plantago pusilla Nutt.
323 323 1900  1 Plantago rugelii Dcne.
323 323 1904  1 Plantago virginica L.

## PLATANACEAE

324 324 1085 1 Platanus occidentalis L.

## POACEAE

325 325 0176A   Aegilops cylindrica Host
325 325 0173  2 Agropyron repens (L.) Beauv.
325 325 0174  2 Agropyron smithii Rydb.
325 325 0175  1 Agropyron trachycaulum (Link) Malte ex
               H.F. Lewis var. trachycaulum
325 325 0176  1 Agropyron trachycaulum (Link) Malte ex
               H.F. Lewis var. unilaterale (Vasey) Malte
325 326 0208  1 Agrostis elliottiana Schultes
325 326 0210  1 Agrostis hiemalis (Walt.) B.S.P.
325 326 0211  1 Agrostis perennans (Walt.) Tuckerman
325 326 0209  1 Agrostis scabra Willd.
325 326 0206  2 Agrostis stolonifera L. var. major
               (Gaudin) Farw.
325 326 0207  2 Agrostis stolonifera L. var. palustris
               (Huds.) Farw.

```
325  327  0198C4   Aira caryophyllea L.
325  327  0214    1 Alopecurus aequalis Sobol.
325  327  0215    1 Alopecurus carolinianus Walt.
325  327  0213    2 Alopecurus pratensis L.
325  327  0204    1 Ammophila breviligulata Fern.
325  328  0333    1 Andropogon elliottii Chapman
325  328  0332    1 Andropogon gerardii Vitman var. gerardii
325  328  0334    1 Andropogon virginicus L.
325  328  0258    2 Anthoxanthum odoratum L.
325  328  0244A     Aristida basiramea Engelm. ex Vasey
325  328  0244    1 Aristida dichotoma Michx.
325  328  0246    1 Aristida longespica Poir. var.
                      geniculata (Raf.) Fern.
325  328  0245    1 Aristida longespica Poir. var. longespica
325  328  0249    1 Aristida oligantha Michx.
325  328  0247    1 Aristida purpurascens Poir.
325  328  0248    1 Aristida ramosissima Engelm. ex Gray
325  328  0243    1 Aristida tuberculosa Nutt.
325  329  0199    2 Arrhenatherum elatius (L.) Beauv. ex J.
                      & C. Presl.
325  329  0119    1 Arundinaria gigantea (Walt.) Muhl.
325  329  0198A     Avena fatua L.
325  329  0198B     Avena sativa L.
325  330  0256    1 Bouteloua curtipendula (Michx.) Torr.
325  330  0235    1 Brachyelytrum erectum (Schreb.) Beauv.
325  330  0124    1 Bromus altissimus Pursh
325  330  0126A   2 Bromus briziformis Fisch. & Mey.
325  330  0123    1 Bromus ciliatus L.
325  330  0129    2 Bromus commutatus Schrad.
325  330  0128    2 Bromus hordeaceus L. ssp. hordeaceus
325  330  0122    2 Bromus inermis Leyss.
325  330  0130    2 Bromus japonicus Thunb. ex Murr.
325  330  0127    1 Bromus kalmii Gray
325  330  0126    2 Bromus secalinus L.
325  330  0126B     Bromus squarrosus L.
325  330  0120    2 Bromus sterilis L.
325  330  0121    2 Bromus tectorum L.
325  332  0202    1 Calamagrostis canadensis (Michx.) Beauv.
325  332  0203    1 Calamagrostis neglecta (Ehrh.) Gaertn.,
                      Mey. & Scherb.
325  332  0205    1 Calamovilfa longifolia (Hook.) Scribn.
325  333  0330    1 Cenchrus incertus M.A. Curtis
325  333  0330B     Cenchrus longispinus (Hack.) Fern.
325  333  0165    1 Chasmanthium latifolium (Michx.) Yates
325  333  0256A     Chloris verticillata Nutt.
325  333  0212    1 Cinna arundinacea L.
325  333  0234A     Crypsis schoenoides (L.) Lam.
325  334  0253    2 Cynodon dactylon (L.) Pers.
```

POACEAE        CONT.

| | | | |
|---|---|---|---|
| 325 | 334 | 0166 | 2 Dactylis glomerata L. |
| 325 | 334 | 0201 | 1 Danthonia spicata (L.) Beauv. ex Roemer & Schultes |
| 325 | 334 | 0198 | 1 Deschampsia cespitosa (L.) Beauv. |
| 325 | 334 | 0164 | 1 Diarrhena americana Beauv. |
| 325 | 334 | 0299 | 1 Dichanthelium acuminatum (Sw.) Gould & Clark var. acuminatum (Sw.) Gould & Clark |
| 325 | 334 | 0295 | 1 Dichanthelium acuminatum (Sw.) Gould var. densiflorum (Rand & Redf.) Gould & Clark |
| 325 | 334 | 0297 | 1 Dichanthelium acuminatum (Sw.) Gould var. implicatum (Scribn.) Gould & Clark |
| 325 | 334 | 0296 | 1 Dichanthelium acuminatum (Sw.) Gould var. lindheimeri (Nash) Gould & Clark |
| 325 | 334 | 0298 | 1 Dichanthelium acuminatum (Sw.) Gould var. villosum (Gray) Gould & Clark |
| 325 | 334 | 0288 | 1 Dichanthelium boreale (Nash) Freckmann |
| 325 | 334 | 0322 | 1 Dichanthelium boscii (Poir.) Gould & Clark |
| 325 | 334 | 0320 | 1 Dichanthelium clandestinum (L.) Gould |
| 325 | 334 | 0319 | 1 Dichanthelium commutatum (Schultes) Gould |
| 325 | 334 | 0309 | 1 Dichanthelium consanguineum (Kunth) Gould & Clark |
| 325 | 334 | 0283 | 1 Dichanthelium depauperatum (Muhl.) Gould |
| 325 | 334 | 0273 | 1 Dichanthelium dichotomum (L.) Gould var. dichotomum |
| 325 | 334 | 0321 | 1 Dichanthelium latifolium (L.) Gould & Clark |
| 325 | 334 | 0287 | 1 Dichanthelium laxiflorum (Lam.) Gould |
| 325 | 334 | 0315 | 1 Dichanthelium leibergii (Vasey) Freckmann |
| 325 | 334 | 0284 | 1 Dichanthelium linearifolium (Scribn.) Gould |
| 325 | 334 | 0316 | 1 Dichanthelium oligosanthes (Schultes) Gould var. oligosanthes |
| 325 | 334 | 0317 | 1 Dichanthelium oligosanthes (Schultes) Gould var. wilcoxianum (Vasey) Gould & Clark |
| 325 | 334 | 0310 | 1 Dichanthelium ovale (Ell.) Gould & Clark var. ovale |
| 325 | 334 | 0311 | 1 Dichanthelium sabulorum (Lam.) Gould & Clark var. patulum (Scribn. & Merr.) Gould & Clark |
| 325 | 334 | 0312 | 1 Dichanthelium sabulorum (Lam.) Gould & Clark var. thinium (A.S. Hitchc. & Chase) Gould & Clark |
| 325 | 334 | 0289 | 1 Dichanthelium sphaerocarpon (Ell.) Gould var. isophyllum (Scribn.) Gould & Clark |
| 325 | 334 | 0314 | 1 Dichanthelium sphaerocarpon (Ell.) Gould var. sphaerocarpon |
| 325 | 337 | 0267A4 | Digitaria arenicola (Swallen) Beetle |

```
325  337  0267   1 Digitaria cognatum (Schultes) Pilger
325  337  0264   1 Digitaria filiformis (L.) Koel.
325  337  0265   2 Digitaria ischaemum (Schreb. ex Schweig)
                   Schreb. ex Muhl.
325  337  0266   2 Digitaria sanguinalis (L.) Scop.
325  338  0251A    Diplachne acuminata Nash
325  338  0251   1 Diplachne panicoides Presl (McNeill)
325  338  0324   1 Echinochloa crusgalli (L.) Beauv.
325  338  0325   1 Echinochloa walteri (Pursh) Heller
325  338  0252   2 Eleusine indica (L.) Gaertn.
325  339  0177   1 Elymus canadensis L.
325  339  0178   1 Elymus riparius Wieg.
325  339  0179   1 Elymus villosus Muhl. ex Willd.
325  339  0183   1 Elymus virginicus L. var. submuticus
                   Hook.
325  339  0181   1 Elymus virginicus L. var. virginicus
325  340  0161   1 Eragrostis capillaris (L.) Nees
325  340  0160   2 Eragrostis cilianensis (All.) E. Mosher
325  340  0162   1 Eragrostis frankii C.A. Mey. ex Steud.
325  340  0158   1 Eragrostis hypnoides (Lam.) B.S.P.
325  340  0158A4   Eragrostis mexicana (Hornem.) Link
325  340  0158B     Eragrostis minor Host
325  340  0163   1 Eragrostis pectinacea (Michx.) Nees
325  340  0159   1 Eragrostis spectabilis (Pursh) Steud.
325  340  0159A    Eragrostis trichodes (Nutt.) Wood
325  340  0330A  1 Erianthus alopecuroides (L.) Ell.
325  341  0136   2 Festuca elatior L.
325  341  0137   1 Festuca obtusa Biehler
325  341  0135   2 Festuca ovina L.
325  341  0138   1 Festuca paradoxa Desv.
325  341  0133   2 Festuca rubra L.
325  341  0134   1 Festuca tenuifolia Sibthorp
325  342  0145   1 Glyceria acutiflora Torr.
325  342  0143   1 Glyceria borealis (Nash) Batchelder
325  342  0141   1 Glyceria canadensis (Michx.) Trin.
325  342  0140   1 Glyceria grandis S. Wats.
325  342  0144   1 Glyceria septentrionalis A.S. Hitchc.
325  342  0139   1 Glyceria striata (Lam.) A.S. Hitchc.
325  342  0255   1 Gymnopogon ambiguus (Michx.) B.S.P.
325  342  0257   1 Hierochloe odorata (L.) Beauv.
325  343  0200   2 Holcus lanatus L.
325  343  0190   2 Hordeum brachyantherum Nevski
325  343  0191   1 Hordeum jubatum L.
325  343  0189   2 Hordeum pusillum Nutt.
325  343  0189A    Hordeum vulgare L.
325  343  0187   1 Hystrix patula Moench
325  343  0188   1 Hystrix patula Moench var. bigeloviana
                   (Fern.) Deam
325  343  0194   1 Koeleria cristata (L.) Pers.
```

| | | | | |
|---|---|---|---|---|
| 325 | 344 | 0262 | 1 | Leersia lenticularis Michx. |
| 325 | 344 | 0261 | 1 | Leersia oryzoides (L.) Sw. |
| 325 | 344 | 0260 | 1 | Leersia virginica Willd. |
| 325 | 344 | 0250 | 1 | Leptochloa filiformis (Lam.) Beauv. |
| 325 | 344 | 0193 | 2 | Lolium multiflorum Lam. |
| 325 | 344 | 0192 | 2 | Lolium perenne L. |
| 325 | 344 | 0168 | 1 | Melica mutica Walt. |
| 325 | 344 | 0169 | 1 | Melica nitens (Scribn.) Nutt. ex Piper |
| 325 | 345 | 0236 | 1 | Milium effusum L. |
| 325 | 345 | 0224A | | Muhlenbergia asperifolia (Nees & Meyen) Parodi |
| 325 | 345 | 0224 | 1 | Muhlenbergia bushii Pohl |
| 325 | 345 | 0217 | 1 | Muhlenbergia capillaris (Lam.) Trin. |
| 325 | 345 | 0220 | 1 | Muhlenbergia cuspidata (Torr.) Rydb. |
| 325 | 345 | 0227A | | Muhlenbergia frondosa (Poir.) Fern. |
| 325 | 345 | 0221 | 1 | Muhlenbergia glabriflora Scribn. |
| 325 | 345 | 0221A | | Muhlenbergia glomerata (Willd.) Trin. |
| 325 | 345 | 0227 | 1 | Muhlenbergia mexicana (L.) Trin. |
| 325 | 345 | 0223 | 1 | Muhlenbergia mexicana (L.) Trin. f. commutata (Scribn.) Wieg. |
| 325 | 345 | 0225 | 1 | Muhlenbergia racemosa (Michx.) B.S.P. |
| 325 | 345 | 0218 | 1 | Muhlenbergia schreberi J.F. Gmel. |
| 325 | 345 | 0219 | 1 | Muhlenbergia sobolifera (Muhl.) Trin. |
| 325 | 345 | 0228 | 1 | Muhlenbergia sylvatica (Torr.) Torr. ex Gray |
| 325 | 345 | 0226 | 1 | Muhlenbergia tenuiflora (Willd.) B.S.P. |
| 325 | 346 | 0238 | 1 | Oryzopsis asperifolia Michx. |
| 325 | 346 | 0237 | 1 | Oryzopsis pungens (Torr. ex Spreng.) A.S. Hitchc. |
| 325 | 346 | 0239 | 1 | Oryzopsis racemosa (Sm.) Ricker |
| 325 | 346 | 0279 | 1 | Panicum anceps Michx. |
| 325 | 346 | 0275 | 1 | Panicum capillare L. |
| 325 | 346 | 0275A4 | | Panicum dichotomiflorum Michx. |
| 325 | 346 | 0274 | 1 | Panicum flexile (Gattinger) Scribn. |
| 325 | 346 | 0277 | 1 | Panicum gattingeri Nash |
| 325 | 346 | 0289A | | Panicum miliaceum L. |
| 325 | 346 | 0276 | 1 | Panicum philadelphicum Bernh. ex.Trin. |
| 325 | 346 | 0281 | 1 | Panicum rigidulum Bosc ex Nees |
| 325 | 346 | 0280 | 1 | Panicum stipitatum Nash |
| 325 | 346 | 0299A1 | | Panicum tuckermanii Fern. |
| 325 | 346 | 0282 | 1 | Panicum verrucosum Muhl. |
| 325 | 346 | 0278 | 1 | Panicum virgatum L. |
| 325 | 347 | 0268 | 1 | Paspalum fluitans (Ell.) Kunth |
| 325 | 347 | 0269 | 1 | Paspalum laeve Michx. var. circulare (Nash) Fern. |
| 325 | 347 | 0270 | 1 | Paspalum pubiflorum Rupr. ex Fourn. var. glabrum Vasey Ex Scribn. |
| 325 | 347 | 0270A | | Paspalum setaceum Michx. var. muhlenbergii (Nash) D. Banks |

| | | | | |
|---|---|---|---|---|
| 325 | 347 | 0271 | 1 | Paspalum setaceum Michx. var. setaceum |
| 325 | 347 | 0270B | | Paspalum setaceum Michx. var. stramineum (Nash) D. Banks |
| 325 | 348 | 0259A | 2 | Phalaris arundinacea L. |
| 325 | 348 | 0259B | | Phalaris canariensis L. |
| 325 | 348 | 0216 | 2 | Phleum pratense L. |
| 325 | 348 | 0167 | 1 | Phragmites australis (Cav.) Trin. ex Steud. |
| 325 | 349 | 0151 | 1 | Poa alsodes Gray |
| 325 | 349 | 0146 | 2 | Poa annua L. |
| 325 | 349 | 0148 | 1 | Poa autumnalis Muhl. ex Ell. |
| 325 | 349 | 0147 | 1 | Poa chapmaniana Scribn. |
| 325 | 349 | 0149 | 2 | Poa compressa L. |
| 325 | 349 | 0157A | 1 | Poa cuspidata Nutt. |
| 325 | 349 | 0150 | 1 | Poa languida A.S. Hitchc. |
| 325 | 349 | 0150A1 | | Poa nemoralis L. |
| 325 | 349 | 0153 | 1 | Poa paludigena Fern. & Wieg. |
| 325 | 349 | 0154 | 1 | Poa palustris L. |
| 325 | 349 | 0155 | 1 | Poa pratensis L. |
| 325 | 349 | 0156 | 1 | Poa sylvestris Gray |
| 325 | 349 | 0152 | 2 | Poa trivialis L. |
| 325 | 349 | 0157 | 1 | Poa wolfii Scribn. |
| 325 | 350 | 0145A | | Puccinellia distans (Jacq.) Parl. |
| 325 | 351 | 0170 | 1 | Schizachne purpurascens (Torr.) Swallen |
| 325 | 351 | 0331 | 1 | Schizachyrium scoparium (Michx.) Nash |
| 325 | 352 | 0176B | | Secale cereale L. |
| 325 | 352 | 0328A | | Setaria faberi Herrm. |
| 325 | 352 | 0326 | 2 | Setaria glauca (L.) Beauv. |
| 325 | 352 | 0328 | 2 | Setaria italica (L.) Beauv. |
| 325 | 352 | 0329 | 2 | Setaria verticillata (L.) Beauv. |
| 325 | 352 | 0327 | 2 | Setaria viridis (L.) Beauv. |
| 325 | 353 | 0336 | 1 | Sorghastrum nutans (L.) Nash |
| 325 | 353 | 0335B | | Sorghum bicolor (L.) Moench ssp. bicolor |
| 325 | 353 | 0335A | 2 | Sorghum bicolor (L.) Moench ssp. drummondii (Steud.) de Wet & Harlan |
| 325 | 353 | 0335 | 1 | Sorghum halepense (L.) Pers. |
| 325 | 353 | 0254 | 1 | Spartina pectinata Link |
| 325 | 353 | 0195 | 1 | Sphenopholis nitida (Biehler) Scribn. |
| 325 | 353 | 0196 | 1 | Sphenopholis obtusata (Michx.) Scribn. var. major (Torr.) K.S. Erdman |
| 325 | 353 | 0197A | 1 | Sphenopholis obtusata (Michx.) Scribn. var. obtusata |
| 325 | 353 | 0234 | 1 | Sporobolus asper (Michx.) Kunth |
| 325 | 353 | 0230 | 1 | Sporobolus clandestinus (Biehler) A.S. Hitchc. |
| 325 | 353 | 0232 | 1 | Sporobolus cryptandrus (Torr.) Gray |
| 325 | 353 | 0233 | 1 | Sporobolus heterolepis (Gray) Gray |
| 325 | 353 | 0231 | 1 | Sporobolus neglectus Nash |
| 325 | 353 | 0229 | 1 | Sporobolus vaginiflorus (Torr. ex Gray) Wood |

POACEAE                CONT.

```
325  354  0240  1 Stipa avenacea L.
325  354  0241  1 Stipa comata Trin. & Rupr.
325  354  0242  1 Stipa spartea Trin.
325  354  0142  1 Torreyochloa pallida (Torr.) Church
325  355  0171  1 Tridens flavus (L.) A.S. Hitchc.
325  355  0172  1 Triplasis purpurea (Walt.) Chapman
325  355  0337  1 Tripsacum dactyloides (L.) L.
325  356  0132  1 Vulpia octoflora (Walt.) Rydb. var.
                  glauca (Nutt.) Fern.
325  356  0131  1 Vulpia octoflora (Walt.) Rydb. var.
                  octoflora
325  356  0336A   Zea mays L.
325  356  0263  1 Zizania aquatica L.
```

POLEMONIACEAE

```
356  357  1689B    Collomia linearis Nutt.
356  358  1690   2 Ipomopsis rubra (L.) Wherry
356  360  1682   1 Phlox amplifolia Britt.
356  360  1689   1 Phlox bifida Beck
356  360  1688   1 Phlox divaricata L.
356  360  1685   1 Phlox glaberrima L. ssp. glaberrima
356  360  1684A4   Phlox glaberrima L. ssp. interior
                   (Wherry) Wherry
356  360  1684   1 Phlox glaberrima L. ssp. triflora
                   (Michx.) Wherry
356  360  1686   1 Phlox maculata L.
356  360  1686A  1 Phlox mollis Wherry
356  360  1683   1 Phlox ovata L.
356  360  1681   1 Phlox paniculata L.
356  360  1687   1 Phlox pilosa L.
356  360  1687A  1 Phlox pilosa L. ssp. fulgida (Wherry)
                   Wherry
356  360  1687B  1 Phlox pilosa L. ssp. pulcherrima Lundell
356  360  1689A  1 Phlox subulata L.
356  362  1691   1 Polemonium reptans L.
```

POLYGALACEAE

```
363  363  1315   1 Polygala cruciata L.
363  363  1308   1 Polygala paucifolia Willd.
363  363  1309   1 Polygala polygama Walt.
363  363  1316   1 Polygala sanguinea L.
363  363  1310   1 Polygala senega L.
363  363  1314   1 Polygala verticillata L. var. ambigua
                   (Nutt.) Wood
363  363  1311   1 Polygala verticillata L. var. isocycla
                   Fern.
```

POLYGALACEAE    CONT.

| 363 | 363 | 1312 | 1 | Polygala verticillata L. var. sphenostachya Pennell |
| 363 | 363 | 1313 | 1 | Polygala verticillata L. var. verticillata |

POLYGONACEAE

| 364 | 371 | 0858 | 2 | Fagopyrum esculentum Moench |
| 364 | 371 | 0859 | 1 | Polygonella articulata (L.) Meisn. |
| 364 | 371 | 0841 | 1 | Polygonum amphibium L. var. emersum Michx. |
| 364 | 371 | 0839 | 1 | Polygonum amphibium L. var. stipulaceum (Coleman) Fern. |
| 364 | 371 | 0853 | 1 | Polygonum arifolium L. var. pubescens (Keller) Fern. |
| 364 | 371 | 0834 | 2 | Polygonum aviculare L. var. vegetum Ledeb. |
| 364 | 371 | 0855A4 | | Polygonum caespitosum Blume var. longisetum (de Bruyn.) A.N. Stewart |
| 364 | 371 | 0846 | 1 | Polygonum careyi Olney |
| 364 | 371 | 0846A4 | | Polygonum cilinode Michx. |
| 364 | 371 | 0855 | 2 | Polygonum convolvulus L. |
| 364 | 371 | 0855A | | Polygonum cuspidatum Sieb. & Zucc. |
| 364 | 371 | 0833 | 1 | Polygonum erectum L. |
| 364 | 371 | 0847 | 1 | Polygonum hydropiper L. |
| 364 | 371 | 0850 | 1 | Polygonum hydropiperoides Michx. |
| 364 | 371 | 0845 | 1 | Polygonum lapathifolium L. |
| 364 | 371 | 0837A | | Polygonum opelousanum Riddell ex Small var. adenocalyx Stanford |
| 364 | 371 | 0851 | 2 | Polygonum orientale L. |
| 364 | 371 | 0842 | 1 | Polygonum pensylvanicum (L.) Small var. pensylvanicum |
| 364 | 371 | 0843 | 1 | Polygonum pensylvanicum (L.) Small var. laevigatum Fern. |
| 364 | 371 | 0849 | 2 | Polygonum persicaria L. |
| 364 | 371 | 0848 | 1 | Polygonum punctatum Ell. |
| 364 | 371 | 0832 | 1 | Polygonum ramosissimum Michx. var. ramosissimum. |
| 364 | 371 | 0854 | 1 | Polygonum sagittatum L. |
| 364 | 371 | 0857A | | Polygonum scandens L. var. cristatum (Engelm. & Gray) Gleason |
| 364 | 371 | 0856 | 1 | Polygonum scandens L. var. dumetorum (L.) Gleason |
| 364 | 371 | 0857 | 1 | Polygonum scandens L. var. scandens |
| 364 | 371 | 0857B | | Polygonum setaceum Baldw. ex Ell. var. interjectum Fern. |
| 364 | 371 | 0857A4 | | Polygonum tenue Michx. |

POLYGONACEAE    CONT.

364  371  0852   1  Polygonum virginianum L.
364  374  0825   2  Rumex acetosella L.
364  374  0826   1  Rumex altissimus Wood
364  374  0830   2  Rumex crispus L.
364  374  0830A1     Rumex hastatulus Baldw. ex Ell.
364  374  0831   2  Rumex obtusifolius L.
364  374  0829   1  Rumex orbiculatus Gray
364  374  0831C1     Rumex patientia L.
364  374  0828   1  Rumex triangulivalvis (Danser) Rech. f.
364  374  0827   1  Rumex verticillatus L.


PÔNTEDERIACEAE

375  375  0598   1  Heteranthera dubia (Jacq.) MacM.
375  375  0597   1  Heteranthera reniformis Ruiz & Pavon
375  375  0596   1  Pontederia cordata L.


PORTULACACEAE

375  375  0893   1  Claytonia virginica L.
375  377  0894   2  Portulaca oleracea L.
375  377  0892   1  Talinum rugospermum Holz.


POTAMOGETONACEAE

377  377  0078   1  Potamogeton amplifolius Tuckerman
377  377  0085   2  Potamogeton crispus L.
377  377  0079   1  Potamogeton diversifolius Raf.
377  377  0089   1  Potamogeton epihydrus Raf.
377  377  0092   1  Potamogeton foliosus Raf. var. foliosus
377  377  0091   1  Potamogeton friesii Rupr.
377  377  0081   1  Potamogeton gramineus L.
377  377  0083   1  Potamogeton illinoensis Morong
377  377  0076   1  Potamogeton natans L.
377  377  0076A     Potamogeton nodosus Poir.
377  377  0097   1  Potamogeton pectinatus L.
377  377  0087   1  Potamogeton praelongus Wulfen
377  377  0082   1  Potamogeton pulcher Tuckerman
377  377  0095   1  Potamogeton pusillus L. var. pusillus
377  377  0093   1  Potamogeton pusillus L. var. tenuissimus
                     Mert. & Koch
377  377  0088   1  Potamogeton richardsonii (Benn.) Rydb.
377  377  0096   1  Potamogeton robbinsii Oakes
377  377  0094   1  Potamogeton strictifolius Benn.
377  377  0094A     Potamogeton vaseyi J.W. Robbins
377  377  0090   1  Potamogeton zosteriformis Fern.

58

## PRIMULACEAE

| | | | | |
|---|---|---|---|---|
| 378 | 378 | 1603 | 2 | Anagallis arvensis L. |
| 378 | 378 | 1604 | 1 | Anagallis minima (L.) Krause |
| 378 | 378 | 1591 | 1 | Androsace occidentalis Pursh |
| 378 | 379 | 1605A5 | | Dodecatheon frenchii (Vasey) Rybd. |
| 378 | 379 | 1605 | 1 | Dodecatheon meadia L. |
| 378 | 380 | 1592 | 1 | Hottonia inflata Ell. |
| 378 | 380 | 1598 | 1 | Lysimachia ciliata L. |
| 378 | 380 | 1600 | 1 | Lysimachia hybrida Michx. |
| 378 | 380 | 1599 | 1 | Lysimachia lanceolata Walt. |
| 378 | 380 | 1594 | 2 | Lysimachia nummularia L. |
| 378 | 380 | 1601 | 1 | Lysimachia quadriflora Sims |
| 378 | 380 | 1596 | 1 | Lysimachia quadrifolia L. |
| 378 | 380 | 1597 | 1 | Lysimachia terrestris (L.) B.S.P. |
| 378 | 380 | 1595 | 1 | Lysimachia thyrsiflora L. |
| 378 | 380 | 1595A | | Lysimachia vulgaris L. |
| 378 | 380 | 1593 | 1 | Samolus valerandi L. ssp. parviflorus (Raf.) Hulten |
| 378 | 380 | 1602 | 1 | Trientalis borealis Raf. |

## RANUNCULACEAE

| | | | | |
|---|---|---|---|---|
| 381 | 381 | 0943A | 1 | Aconitum uncinatum L. |
| 381 | 381 | 0938A | | Actaea pachypoda Ell. |
| 381 | 381 | 0939 | 1 | Actaea rubra (Ait.) Willd. |
| 381 | 381 | 0946 | 1 | Anemone canadensis L. |
| 381 | 381 | 0945 | 1 | Anemone caroliniana Walt. |
| 381 | 381 | 0947 | 1 | Anemone cylindrica Gray |
| 381 | 381 | 0944 | 1 | Anemone quinquefolia L. var. interior Fern. |
| 381 | 381 | 0948 | 1 | Anemone virginiana L. |
| 381 | 382 | 0941 | 1 | Aquilegia canadensis L. |
| 381 | 382 | 0935 | 1 | Caltha palustris L. |
| 381 | 382 | 0940 | 1 | Cimicifuga racemosa (L.) Nutt. |
| 381 | 382 | 0953 | 1 | Clematis pitcheri Torr. & Gray |
| 381 | 382 | 0953A4 | | Clematis terniflora DC. |
| 381 | 382 | 0952 | 1 | Clematis viorna L. |
| 381 | 382 | 0954 | 1 | Clematis virginiana L. |
| 381 | 383 | 0942 | 2 | Consolida ambigua (L.) Ball & Heywood |
| 381 | 383 | 0937 | 1 | Coptis trifolia (L.) Salisb. ssp. groenlandica (Oeder) Hulten |
| 381 | 383 | 0943 | 1 | Delphinium tricorne Michx. |
| 381 | 384 | 0950 | 1 | Hepatica nobilis P. Mill. var. acuta (Pursh) Steyermark |
| 381 | 384 | 0951 | 1 | Hepatica nobilis P. Mill. var. obtusa (Pursh) Steyermark |
| 381 | 385 | 0934 | 1 | Hydrastis canadensis L. |

## RANUNCULACEAE     CONT.

| | | | | |
|---|---|---|---|---|
| 381 | 385 | 0936 | 1 | Isopyrum biternatum (Raf.) Torr. & Gray |
| 381 | 385 | 0955 | 2 | Myosurus minimus L. |
| 381 | 385 | 0963 | 1 | Ranunculus abortivus L. |
| 381 | 385 | 0968 | 2 | Ranunculus acris L. |
| 381 | 385 | 0962 | 1 | Ranunculus ambigens S. Wats. |
| 381 | 385 | 0958 | 1 | Ranunculus aquatilis L. var. capillaceus (Thuill.) DC. |
| 381 | 385 | 0966 | 2 | Ranunculus bulbosus L. |
| 381 | 385 | 0966A1 | | Ranunculus carolinianus DC. |
| 381 | 385 | 0969 | 1 | Ranunculus fascicularis Muhl. ex Bigelow |
| 381 | 385 | 0957 | 1 | Ranunculus flabellaris Raf. |
| 381 | 385 | 0972 | 1 | Ranunculus hispidus Michx. |
| 381 | 385 | 0959 | 1 | Ranunculus longirostre Godr. |
| 381 | 385 | 0965 | 1 | Ranunculus micranthus (Gray) Nutt. ex Torr. & Gray |
| 381 | 385 | 0970 | 1 | Ranunculus pensylvanicus L. f. |
| 381 | 385 | 0960 | 1 | Ranunculus pusillus Poir. |
| 381 | 385 | 0967 | 1 | Ranunculus recurvatus Poir. |
| 381 | 385 | 0974A | | Ranunculus repens L. |
| 381 | 385 | 0971 | 2 | Ranunculus repens L. var. repens |
| 381 | 385 | 0964 | 1 | Ranunculus sceleratus L. |
| 381 | 385 | 0973 | 1 | Ranunculus septentrionalis Poir. |
| 381 | 388 | 0977 | 1 | Thalictrum dasycarpum Fisch. & Lall. |
| 381 | 388 | 0977A4 | | Thalictrum dasycarpum Fisch. & Lall. var. hypoglaucum (Rydb.) Boivin |
| 381 | 388 | 0975 | 1 | Thalictrum dioicum L. |
| 381 | 388 | 0978 | 1 | Thalictrum pubescens Pursh var. pubescens |
| 381 | 388 | 0976 | 1 | Thalictrum revolutum DC. |
| 381 | 388 | 0949 | 1 | Thalictrum thalictroides (L.) Eames & Boivin |
| 381 | 388 | 0956 | 1 | Trautvetteria caroliniensis (Walt.) Vail |

## RHAMNACEAE

| | | | | |
|---|---|---|---|---|
| 388 | 388 | 1376 | 1 | Ceanothus americanus L. |
| 388 | 388 | 1377 | 1 | Ceanothus herbaceus Raf. |
| 388 | 390 | 1375 | 1 | Rhamnus alnifolia L'Her |
| 388 | 390 | 1373 | 1 | Rhamnus caroliniana Walt. |
| 388 | 390 | 1373A | 1 | Rhamnus caroliniana Walt. var. mollis Fern. |
| 388 | 390 | 1373B | | Rhamnus cathartica L. |
| 388 | 390 | 1375A | 2 | Rhamnus frangula L. |
| 388 | 390 | 1374 | 1 | Rhamnus lanceolata Pursh |
| 388 | 390 | 1374A1 | | Rhamnus lanceolata Pursh var. glabrata Gleason |

## ROSACEAE

| | | | | |
|---|---|---|---|---|
| 391 | 391 | 1168 | 1 | Agrimonia gryposepala Wallr. |
| 391 | 391 | 1171 | 1 | Agrimonia parviflora Ait. |
| 391 | 391 | 1170 | 1 | Agrimonia pubescens Wallr. |
| 391 | 391 | 1169 | 1 | Agrimonia rostellata Wallr. |
| 391 | 391 | 1099C | | Amelanchier arborea (Michx. f.) Fern. |
| 391 | 391 | 1100 | 1 | Amelanchier arborea (Michx. f.) Fern. var. laevis (Wieg.) Ahles |
| 391 | 391 | 1099 | 1 | Amelanchier canadensis (L.) Medic. |
| 391 | 391 | 1099A | 1 | Amelanchier canadensis X humilis |
| 391 | 391 | 1099B | 1 | Amelanchier canadensis X laevis |
| 391 | 391 | 1098 | 1 | Amelanchier humilis Wieg. |
| 391 | 391 | 1098A | 1 | Amelanchier humilis X laevis |
| 391 | 391 | 1100A | | Amelanchier sanguinea (Pursh) DC. |
| 391 | 392 | 1096 | 1 | Aronia melanocarpa (Michx.) Ell. |
| 391 | 392 | 1097 | 1 | Aronia prunifolia (Marsh.) Rehd. |
| 391 | 392 | 1090 | 1 | Aruncus dioicus (Walt.) Fern. |
| 391 | 393 | 1113 | 1 | Crataegus biltmoreana Beadle |
| 391 | 393 | 1125 | 1 | Crataegus calpodendron (Ehrh.) Medic. |
| 391 | 393 | 1121 | 1 | Crataegus chrysocarpa Ashe |
| 391 | 393 | 1101 | 1 | Crataegus crus-galli L. |
| 391 | 393 | 1108 | 1 | Crataegus disperma Ashe |
| 391 | 393 | 1107 | 1 | Crataegus grandis Ashe |
| 391 | 393 | 1111 | 1 | Crataegus intricata Lange |
| 391 | 393 | 1123 | 1 | Crataegus kelloggii Sarg. |
| 391 | 393 | 1114 | 1 | Crataegus macrosperma Ashe |
| 391 | 393 | 1110 | 1 | Crataegus margaretta Ashe |
| 391 | 393 | 1122 | 1 | Crataegus mollis (Torr. & Gray) Scheele |
| 391 | 393 | 1120 | 1 | Crataegus pedicellata Sarg. |
| 391 | 393 | 1124 | 1 | Crataegus phaenopyrum (L. f.) Medic. |
| 391 | 393 | 1119 | 1 | Crataegus prona Ashe |
| 391 | 393 | 1115 | 1 | Crataegus pruinosa (Wendl. f.) K. Koch |
| 391 | 393 | 1105 | 1 | Crataegus punctata Jacq. |
| 391 | 393 | 1126 | 1 | Crataegus succulenta Schrad. ex Link |
| 391 | 393 | 1109 | 1 | Crataegus viridis L. |
| 391 | 393 | 1127 | 1 | Crataegus X incaedua Sarg. |
| 391 | 397 | 1149 | 2 | Duchesnea indica (Andr.) Focke |
| 391 | 398 | 1167 | 1 | Filipendula rubra (Hill) B.L. Robins. |
| 391 | 398 | 1148 | 1 | Fragaria vesca L. |
| 391 | 398 | 1148A | | Fragaria vesca L. ssp. americana (Porter) Staudt |
| 391 | 398 | 1146 | 1 | Fragaria virginiana Duchesne |
| 391 | 398 | 1147 | 1 | Fragaria virginiana Duchesne ssp. platypetala (Rydb.) Staudt |
| 391 | 398 | 1164 | 1 | Geum aleppicum Jacq. |
| 391 | 398 | 1161 | 1 | Geum canadense Jacq. |
| 391 | 398 | 1162 | 1 | Geum canadense Jacq. var. grimesii Fern. & Weath. |
| 391 | 398 | 1165 | 1 | Geum laciniatum Murr. var. laciniatum |
| 391 | 398 | 1166 | 1 | Geum laciniatum Murr. var. trichocarpum Fern. |

| | | | | |
|---|---|---|---|---|
| 391 | 398 | 1160 | 1 | Geum rivale L. |
| 391 | 398 | 1159 | 1 | Geum vernum (Raf.) Torr. & Gray |
| 391 | 398 | 1163 | 1 | Geum virginianum L. |
| 391 | 400 | 1092 | 1 | Malus coronaria (L.) P. Mill. var. coronaria |
| 391 | 400 | 1093 | 1 | Malus coronaria (L.) P. Mill. var. dasycalyx Rehd. |
| 391 | 400 | 1094 | 1 | Malus ioensis (Wood) Britt. |
| 391 | 400 | 1091F | | Malus sylvestris (L.) P. Mill. |
| 391 | 400 | 1086 | 1 | Physocarpus opulifolius (L.) Maxim. |
| 391 | 400 | 1087 | 1 | Physocarpus opulifolius (L.) Maxim. var. intermedius (Rydb.) B.L. Robins. |
| 391 | 400 | 1091 | 1 | Porteranthus stipulatus (Muhl. ex Willd.) Britt. |
| 391 | 400 | 1152 | 1 | Potentilla anserina L. |
| 391 | 400 | 1156 | 1 | Potentilla argentea L. |
| 391 | 400 | 1151 | 1 | Potentilla arguta Pursh |
| 391 | 400 | 1150 | 1 | Potentilla fruticosa L. |
| 391 | 400 | 1155A | | Potentilla norvegica L. |
| 391 | 400 | 1155 | 1 | Potentilla norvegica L. ssp. monspeliensis (L.) Aschers & Graebn. |
| 391 | 400 | 1153 | 1 | Potentilla palustris (L.) Scop. |
| 391 | 400 | 1154 | 2 | Potentilla recta L. |
| 391 | 400 | 1157 | 1 | Potentilla simplex Michx. var. simplex |
| 391 | 403 | 1182 | 1 | Prunus americana Marsh. |
| 391 | 403 | 1185 | 2 | Prunus angustifolia Marsh. |
| 391 | 403 | 1185A | | Prunus avium (L.) L. |
| 391 | 403 | 1187 | 1 | Prunus hortulana Bailey |
| 391 | 403 | 1191 | 2 | Prunus mahaleb L. |
| 391 | 403 | 1183 | 1 | Prunus nigra Ait. |
| 391 | 403 | 1188 | 1 | Prunus pensylvanica L. f. |
| 391 | 403 | 1188A | | Prunus persica (L.) Batsch |
| 391 | 403 | 1184 | 1 | Prunus pumila L. |
| 391 | 403 | 1190 | 1 | Prunus serotina Ehrh. |
| 391 | 403 | 1189 | 1 | Prunus virginiana L. |
| 391 | 403 | 1189A | 1 | Prunus virginiana L. var. demissa (Nutt.) Torr. |
| 391 | 404 | 1091A | | Pyrus communis L. |
| 391 | 404 | 1180 | 1 | Rosa arkansana Porter var. suffulta (Greene) Cockerell |
| 391 | 404 | 1179 | 1 | Rosa blanda Ait. var. blanda |
| 391 | 404 | 1179A | 1 | Rosa blanda Ait. var. carpohispida Schuette |
| 391 | 404 | 1179B | 1 | Rosa blanda Ait. var. glandulosa Schuette |
| 391 | 404 | 1179C | | Rosa canina L. |
| 391 | 404 | 1177 | 1 | Rosa carolina L. var. carolina |
| 391 | 404 | 1176C | 1 | Rosa carolina L. var. deamii (Erlanson) Deam |
| 391 | 404 | 1176B | 1 | Rosa carolina L. var. sabulosa Erlanson |

ROSACEAE          CONT.

| | | | | |
|---|---|---|---|---|
| 391 | 404 | 1178 | 1 | Rosa carolina L. var. villosa (Best) Rehd. |
| 391 | 404 | 1175 | 2 | Rosa eglanteria L. |
| 391 | 404 | 1178B | | Rosa gallica L. |
| 391 | 404 | 1175A | 2 | Rosa micrantha Borrer  Ex. Sm. |
| 391 | 404 | 1175B | | Rosa multiflora Thunb. ex Murr. |
| 391 | 404 | 1176 | 1 | Rosa palustris Marsh. |
| 391 | 404 | 1181 | 1 | Rosa rudiuscula Greene |
| 391 | 404 | 1173 | 1 | Rosa setigera Michx. var. setigera |
| 391 | 404 | 1174 | 1 | Rosa setigera Michx. var. tomentosa Torr. & Gray |
| 391 | 405 | 1145 | 1 | Rubus abactus Bailey |
| 391 | 405 | 1138 | 1 | Rubus allegheniensis Porter ex Bailey |
| 391 | 405 | 1139 | 1 | Rubus alumnus Bailey |
| 391 | 405 | 1141 | 1 | Rubus argutus Link |
| 391 | 405 | 1136 | 1 | Rubus centralis Bailey |
| 391 | 405 | 1137 | 1 | Rubus deamii Bailey |
| 391 | 405 | 1135 | 1 | Rubus enslenii Tratt. |
| 391 | 405 | 1134 | 1 | Rubus flagellaris Willd. |
| 391 | 405 | 1144 | 1 | Rubus frondosus Bigelow |
| 391 | 405 | 1133 | 1 | Rubus hispidus L. |
| 391 | 405 | 1131 | 1 | Rubus idaeus L. ssp. sachalinensis (Levl.) Focke |
| 391 | 405 | 1143 | 1 | Rubus impar Bailey |
| 391 | 405 | 1140A | 4 | Rubus laciniatus Willd. |
| 391 | 405 | 1140 | 1 | Rubus laudatus Berger |
| 391 | 405 | 1130 | 1 | Rubus occidentalis L. |
| 391 | 405 | 1128 | 1 | Rubus odoratus L. |
| 391 | 405 | 1142 | 1 | Rubus ostryifolius Rydb. |
| 391 | 405 | 1142A | | Rubus pensilvanicus Poir. |
| 391 | 405 | 1130A | 2 | Rubus phoenicolasius Maxim. |
| 391 | 405 | 1129 | 1 | Rubus pubescens Raf. |
| 391 | 410 | 1172 | 1 | Sanguisorba canadensis L. |
| 391 | 410 | 1095A | | Sorbus aucuparia L. |
| 391 | 410 | 1095 | 1 | Sorbus decora (Sarg.) Schneid. |
| 391 | 410 | 1088 | 1 | Spiraea alba Du Roi. |
| 391 | 410 | 1089 | 1 | Spiraea tomentosa L. |
| 391 | 411 | 1158 | 1 | Waldsteinia fragarioides (Michx.) Tratt. |

RUBIACEAE

| | | | | |
|---|---|---|---|---|
| 411 | 411 | 1910 | 1 | Cephalanthus occidentalis L. |
| 411 | 412 | 1913 | 1 | Diodia teres Walt. |
| 411 | 412 | 1922 | 1 | Galium aparine L. |
| 411 | 412 | 1927 | 1 | Galium asprellum Michx. |
| 411 | 412 | 1919 | 1 | Galium boreale L. |
| 411 | 412 | 1921A | | Galium brevipes Fern. & Wieg. |
| 411 | 412 | 1915 | 1 | Galium circaezans Michx. var. circaezans |

63

## RUBIACEAE　　　CONT.

| 411 | 412 | 1916 | 1 | Galium circaezans Michx. var. hypomalacum Fern. |
| 411 | 412 | 1926 | 1 | Galium concinnum Torr. & Gray |
| 411 | 412 | 1928 | 1 | Galium labradoricum (Wieg.) Wieg. |
| 411 | 412 | 1917 | 1 | Galium lanceolatum Torr. |
| 411 | 412 | 1917A | | Galium mollugo L. |
| 411 | 412 | 1924 | 1 | Galium obtusum Bigelow |
| 411 | 412 | 1925 | 2 | Galium parisiense L. |
| 411 | 412 | 1918 | 1 | Galium pilosum Ait. |
| 411 | 412 | 1929 | 1 | Galium tinctorium L. |
| 411 | 412 | 1930 | 1 | Galium trifidum L. |
| 411 | 412 | 1923 | 1 | Galium triflorum Michx. |
| 411 | 415 | 1906 | 1 | Hedyotis caerulea (L.) Hook. |
| 411 | 415 | 1909 | 1 | Hedyotis longifolia (Gaertn.) Hook. |
| 411 | 415 | 1908 | 1 | Hedyotis nigricans (Lam.) Fosberg |
| 411 | 415 | 1907 | 1 | Hedyotis purpurea (L.) Torr. & Gray |
| 411 | 417 | 1912 | 1 | Mitchella repens L. |
| 411 | 418 | 1914 | 1 | Spermacoce glabra Michx. |

## RUTACEAE

| 418 | 420 | 1306 | 1 | Ptelea trifoliata L. ssp. trifoliata |
| 418 | 420 | 1306A4 | | Ptelea trifoliata L. ssp. trifoliata var. mollis Torr. & Gray |
| 418 | 420 | 1305 | 1 | Zanthoxylum americanum P. Mill. |

## SALICACEAE

| 420 | 420 | 0728A | 2 | Populus alba L. |
| 420 | 420 | 0729A | 1 | Populus balsamifera L. |
| 420 | 420 | 0730 | 1 | Populus deltoides Bartr. ex Marsh. |
| 420 | 420 | 0731 | 1 | Populus grandidentata Michx. |
| 420 | 420 | 0729 | 1 | Populus heterophylla L. |
| 420 | 420 | 0729C | | Populus nigra L. |
| 420 | 420 | 0732 | 1 | Populus tremuloides Michx. |
| 420 | 420 | 0729B | 1 | Populus X canescens (Ait.) Sm. |
| 420 | 420 | 0729D1 | | Populus X jackii Sarg. |
| 420 | 421 | 0739 | 1 | Salix alba L. |
| 420 | 421 | 0734 | 1 | Salix amygdaloides Anderss. |
| 420 | 421 | 0749 | 1 | Salix bebbiana Sarg. |
| 420 | 421 | 0751 | 1 | Salix candida Flugge ex Willd. |
| 420 | 421 | 0738 | 1 | Salix caroliniana Michx. |
| 420 | 421 | 0752 | 1 | Salix cordata Muhl. |
| 420 | 421 | 0744 | 1 | Salix discolor Muhl. |
| 420 | 421 | 0742 | 1 | Salix exigua Nutt. |
| 420 | 421 | 0740 | 1 | Salix fragilis L. |

SALICACEAE          CONT.

| | | | | |
|---|---|---|---|---|
| 420 | 421 | 0747 | 1 | Salix humilis Marsh. |
| 420 | 421 | 0748 | 1 | Salix humilis Marsh. var. microphylla (Anderss.) Fern. |
| 420 | 421 | 0736 | 1 | Salix lucida Muhl. |
| 420 | 421 | 0754 | 1 | Salix myricoides Muhl. var. myricoides |
| 420 | 421 | 0733 | 1 | Salix nigra Marsh. |
| 420 | 421 | 0750 | 1 | Salix pedicellaris Pursh |
| 420 | 421 | 0750A | | Salix pentandra L. |
| 420 | 421 | 0745 | 1 | Salix petiolaris Sm. |
| 420 | 421 | 0745A | | Salix rigida Muhl. |
| 420 | 421 | 0746 | 1 | Salix sericea Marsh. |
| 420 | 421 | 0735 | 1 | Salix serissima (Bailey) Fern. |

SANTALACEAE

| | | | | |
|---|---|---|---|---|
| 428 | 428 | 0820 | 1 | Comandra umbellata (L.) Nutt. ssp. umbellata |

SAPINDACEAE

| | | | | |
|---|---|---|---|---|
| 429 | 429 | 1370A | 2 | Koelreuteria paniculata Laxm. |

SAPOTACEAE

| | | | | |
|---|---|---|---|---|
| 430 | 430 | 1606 | 1 | Bumelia lycioides (L.) Pers. |

SARRACENIACEAE

| | | | | |
|---|---|---|---|---|
| 431 | 431 | 1058 | 1 | Sarracenia purpurea L. |

SAURURACEAE

| | | | | |
|---|---|---|---|---|
| 431 | 431 | 0728 | 1 | Saururus cernuus L. |

SAXIFRAGACEAE

| | | | | |
|---|---|---|---|---|
| 431 | 431 | 1076 | 1 | Chrysosplenium americanum Schwein. ex Hook. |
| 431 | 432 | 1068 | 1 | Heuchera americana L. var. americana |
| 431 | 432 | 1069 | 1 | Heuchera americana L. var. hirsuticaulis (Wheelock) Rhosendahl, Butters & Lakela |
| 431 | 432 | 1074 | 1 | Heuchera parviflora Nutt. ex Torr. & Gray |
| 431 | 432 | 1071 | 1 | Heuchera richardsonii R. Br. |

## SAXIFRAGACEAE    CONT.

```
431  432  1073    1 Heuchera villosa Michx.
431  433  1078    1 Hydrangea arborescens L.
431  433  1075    1 Mitella diphylla L.
431  434  1077    1 Parnassia glauca Raf.
431  434  1064    1 Penthorum sedoides L.
431  434  1077A     Philadelphus coronarius L.
431  434  1080A4    Ribes americanum P. Mill.
431  434  1080    1 Ribes cynosbati L.
431  434  1082    1 Ribes hirtellum Michx.
431  434  1081    1 Ribes missouriense Nutt. ex Torr. & Gray
431  434  1079C     Ribes odoratum H. Wendl.
431  436  1067A3    Saxifraga forbesii Vasey
431  436  1067    1 Saxifraga pensylvanica L.
431  436  1066    1 Saxifraga virginiensis Michx.
431  438  1065    1 Sullivantia sullivantii (Torr. & Gray)
                    Britt.
```

## SCHEUCHZERIACEAE

```
438  438  0104    1 Scheuchzeria palustris L. ssp. americana
                    (Fern.) Hulten
438  438  0102    1 Triglochin maritima L.
438  438  0103    1 Triglochin palustre L.
```

## SCROPHULARIACEAE

```
438  438  1857    1 Agalinis besseyana Britt.
438  438  1860    1 Agalinis gattingeri (Small) Small
438  438  1854    1 Agalinis paupercula (Gray) Britt.
438  438  1855    1 Agalinis paupercula (Gray) Britt. var.
                    borealis Pennell
438  438  1853    1 Agalinis purpurea (L.) Pennell
438  438  1859    1 Agalinis skinneriana (Wood) Britt.
438  438  1858    1 Agalinis tenuifolia (Vahl) Raf. var.
                    parviflora (Nutt.) Pennell
438  438  1856    1 Agalinis tenuifolia (Vahl) Raf. var.
                    tenuifolia
438  439  1862    1 Aureolaria flava (L.) Farw. var. flava
438  439  1863    1 Aureolaria flava (L.) Farw. var.
                    macrantha Pennell
438  439  1865    1 Aureolaria grandiflora (Benth.) Pennell
                    var. pulchra Pennell
438  439  1867    1 Aureolaria pedicularia (L.) Raf. var.
                    ambigens (Fern.) Farw.
438  439  1866A  1 Aureolaria pedicularia (L.) Raf. var.
                    intercedens Pennell
438  439  1866    1 Aureolaria pedicularia (L.) Raf. var.
                    pedicularia
```

SCROPHULARIACEAE CONT.

```
438  439   1864   1 Aureolaria virginica (L.) Pennell
438  440   1835   1 Bacopa rotundifolia (Michx.) Wettst.
438  440   1851   1 Besseya bullii (Eat.) Rydb.
438  440   1868   1 Buchnera americana L.
438  440   1869   1 Castilleja coccinea (L.) Spreng.
438  442   1815   2 Chaenorrhinum minus (L.) Lange
438  442   1821A  1 Chelone glabra L. var. elatior Raf.
438  442   1820   1 Chelone glabra L. var. glabra
438  442   1821   1 Chelone glabra L. var. linifolia Coleman
438  442   1819   1 Chelone obliqua L. var. speciosa Pennell
                     & Wherry
438  442   1816   1 Collinsia verna Nutt.
438  443   1852   1 Dasystoma macrophylla (Nutt.) Raf.
438  444   1832   1 Gratiola neglecta Torr.
438  444   1833   1 Gratiola virginiana L.
438  444   1812   2 Kickxia elatine (L.) Dumort
438  444   1834   1 Leucospora multifida (Michx.) Nutt.
438  444   1814   1 Linaria canadensis (L.) Dum.-Cours.
438  444   1814A1   Linaria genistifolia (L.) P. Mill. ssp.
                     dalmatica (L.) Maire & Petitmengin
438  444   1813   2 Linaria vulgaris P. Mill.
438  445   1838   1 Lindernia dubia (L.) Pennell var.
                     anagallidea (Michx.) Cooperrider
438  445   1836   1 Lindernia dubia (L.) Pennell var. dubia
438  445   1870   1 Melampyrum lineare Desr. var. latifolium
                     Bart.
438  445   1871   1 Melampyrum lineare Desr. var. pectinatum
                     (Pennell) Fern.
438  445   1831   1 Mimulus alatus Ait.
438  445   1830   1 Mimulus ringens L.
438  449   1872   1 Pedicularis canadensis L.
438  449   1873   1 Pedicularis lanceolata Michx.
438  450   1825   1 Penstemon alluviorum Pennell
438  450   1823   1 Penstemon calycosus Small
438  450   1827   1 Penstemon canescens (Britt.) Britt.
438  450   1826   1 Penstemon deamii Pennell
438  450   1824   1 Penstemon digitalis Nutt.
438  450   1824A    Penstemon gracilis Nutt.
438  450   1829   1 Penstemon hirsutus (L.) Willd.
438  450   1828   1 Penstemon pallidus Small
438  450   1822   1 Penstemon tubiflorus Nutt.
438  454   1818   1 Scrophularia lanceolata Pursh
438  454   1817   1 Scrophularia marilandica L.
438  454   1861   1 Tomanthera auriculata (Michx.) Raf.
438  454   1809   2 Verbascum blattaria L.
438  454   1810   2 Verbascum phlomoides L.
438  454   1811   2 Verbascum thapsus L.
438  454   1847   1 Veronica americana (Raf.) Schwein. ex
                     Benth.
```

67

## SCROPHULARIACEAE CONT.

| | | | | |
|---|---|---|---|---|
| 438 | 454 | 1842 | 2 | Veronica arvensis L. |
| 438 | 454 | 1845 | 2 | Veronica catenata Pennell |
| 438 | 454 | 1848 | 1 | Veronica cymbalaria Bodard |
| 438 | 454 | 1844 | 2 | Veronica officinalis L. |
| 438 | 454 | 1839 | 1 | Veronica peregrina L. ssp. peregrina |
| 438 | 454 | 1840 | 1 | Veronica peregrina L. ssp. xalapensis (H.B.K.) Pennell |
| 438 | 454 | 1843 | 2 | Veronica persica Poir. |
| 438 | 454 | 1843A | | Veronica polita Fries |
| 438 | 454 | 1846 | 1 | Veronica scutellata L. |
| 438 | 454 | 1841 | 2 | Veronica serpyllifolia L. |
| 438 | 454 | 1850 | 1 | Veronicastrum virginicum (L.) Farw. |

## SIMAROUBACEAE

| | | | | |
|---|---|---|---|---|
| 455 | 455 | 1307 | 2 | Ailanthus altissima (P. Mill.) Swingle |

## SMILACACEAE

| | | | | |
|---|---|---|---|---|
| 455 | 455 | 0671 | 1 | Smilax bona-nox L. |
| 455 | 455 | 0669 | 1 | Smilax ecirrhata (Engelm. ex Kunth) S. Wats. |
| 455 | 455 | 0670 | 1 | Smilax glauca Walt. var. glauca |
| 455 | 455 | 0667 | 1 | Smilax herbacea L. |
| 455 | 455 | 0673 | 1 | Smilax hispida Muhl. |
| 455 | 455 | 0668 | 1 | Smilax lasioneuron Hook. |
| 455 | 455 | 0666 | 1 | Smilax pulverulenta Michx. |
| 455 | 455 | 0672 | 1 | Smilax rotundifolia L. |

## SOLANACEAE

| | | | | |
|---|---|---|---|---|
| 456 | 456 | 1808 | 2 | Datura stramonium L. |
| 456 | 457 | 1796 | 2 | Lycium barbarum L. |
| 456 | 457 | 1807A | | Lycopersicon esculentum P. Mill. |
| 456 | 457 | 1795 | 2 | Nicandra physalodes (L.) Gaertn. |
| 456 | 457 | 1808A | | Petunia violacea Lindl. |
| 456 | 457 | 1799 | 1 | Physalis heterophylla Nees var. heterophylla |
| 456 | 457 | 1802A | | Physalis ixocarpa Brot. ex Hornem. |
| 456 | 457 | 1800 | 1 | Physalis pubescens L. |
| 456 | 457 | 1801 | 1 | Physalis pubescens L. var. integrifolia (Dunal) Waterfall |
| 456 | 457 | 1797 | 1 | Physalis subglabrata Mackenzie & Bush |
| 456 | 457 | 1798 | 1 | Physalis virginiana P. Mill. |
| 456 | 458 | 1804A | | Solanum americanum P. Mill. |
| 456 | 458 | 1804 | 1 | Solanum carolinense L. |

SOLANACEAE        CONT.

| | | | | |
|---|---|---|---|---|
| 456 | 458 | 1805 | 2 | Solanum cornutum Lam. |
| 456 | 458 | 1807 | 1 | Solanum dulcamara L. |
| 456 | 458 | 1806 | 1 | Solanum nigrum L. |

SPARGANIACEAE

| | | | | |
|---|---|---|---|---|
| 459 | 459 | 0073 | 1 | Sparganium americanum Nutt. |
| 459 | 459 | 0072 | 1 | Sparganium androcladum (Engelm.) Morong |
| 459 | 459 | 0074 | 1 | Sparganium chlorocarpum Rydb. |
| 459 | 459 | 0071 | 1 | Sparganium eurycarpum Engelm. |

STAPHYLEACEAE

| | | | | |
|---|---|---|---|---|
| 460 | 460 | 1361 | 1 | Staphylea trifolia L. |

STYRACACEAE

| | | | | |
|---|---|---|---|---|
| 460 | 461 | 1608 | 1 | Styrax americana Lam. |

THYMELIACEAE

| | | | | |
|---|---|---|---|---|
| 462 | 462 | 1459 | 1 | Dirca palustris L. |

TILIACEAE

| | | | | |
|---|---|---|---|---|
| 462 | 462 | 1387 | 1 | Tilia americana L. |
| 462 | 462 | 1388 | 1 | Tilia heterophylla Vent. |

TYPHACEAE

| | | | | |
|---|---|---|---|---|
| 463 | 463 | 0070 | 1 | Typha angustifolia L. |
| 463 | 463 | 0069 | 1 | Typha latifolia L. |

ULMACEAE

| | | | | |
|---|---|---|---|---|
| 463 | 463 | 0804 | 1 | Celtis laevigata Willd. |
| 463 | 463 | 0802 | 1 | Celtis occidentalis L. var. canina (Raf.) Sarg. |
| 463 | 463 | 0803 | 1 | Celtis occidentalis L. var. occidentalis |
| 463 | 463 | 0805 | 1 | Celtis occidentalis L. var. pumila (Pursh) Gray |
| 463 | 463 | 0805A | | Celtis tenuifolia Nutt. |

## ULMACEAE CONT.

| | | | | |
|---|---|---|---|---|
| 463 | 463 | 0800 | 1 | Ulmus alata Michx. |
| 463 | 463 | 0801 | 1 | Ulmus americana L. |
| 463 | 463 | 0798A | | Ulmus pumila L. |
| 463 | 463 | 0798 | 1 | Ulmus rubra Muhl. |
| 463 | 463 | 0799 | 1 | Ulmus thomasii Sarg. |

## URTICACEAE

| | | | | |
|---|---|---|---|---|
| 463 | 463 | 0816 | 1 | Boehmeria cylindrica (L.) Sw. var. cylindrica |
| 463 | 463 | 0817 | 1 | Boehmeria cylindrica (L.) Sw. var. drummondiana (Weddell) Weddell |
| 463 | 464 | 0813 | 1 | Laportea canadensis (L.) Weddell |
| 463 | 464 | 0818 | 1 | Parietaria pensylvanica Muhl. ex Willd. |
| 463 | 464 | 0815 | 1 | Pilea fontana (Lunell) Rydb. |
| 463 | 464 | 0814 | 1 | Pilea pumila (L.) Gray |
| 463 | 464 | 0811A | 2 | Urtica dioica L. |
| 463 | 464 | 0812 | 1 | Urtica dioica L. ssp. gracilis (Ait.) Seland. var. procera (Muhl.) Weddell |

## VALERIANACEAE

| | | | | |
|---|---|---|---|---|
| 465 | 465 | 1960 | 1 | Valeriana edulis Nutt. ex Torr. & Gray |
| 465 | 465 | 1960A | | Valeriana edulis Nutt. ex Torr. & Gray ssp. ciliata (Torr. & Gray) F.G. Mey. |
| 465 | 465 | 1960B | | Valeriana officinalis L. |
| 465 | 465 | 1959 | 1 | Valeriana pauciflora Michx. |
| 465 | 465 | 1959A1 | | Valeriana sitchensis Bong. ssp. uliginosa (Torr. & Gray) F.G. Mey. |
| 465 | 465 | 1957 | 1 | Valerianella chenopodiifolia (Pursh) DC. |
| 465 | 465 | 1956A | 2 | Valerianella locusta (L.) Betcke |
| 465 | 465 | 1958 | 1 | Valerianella umbilicata (Sullivant) Wood |

## VERBENACEAE

| | | | | |
|---|---|---|---|---|
| 465 | 466 | 1724B | | Glandularia bipinnatifida (Nutt.) Nutt. |
| 465 | 466 | 1718 | 2 | Glandularia canadensis (L.) Nutt. |
| 465 | 467 | 1897 | 1 | Phryma leptostachya L. |
| 465 | 467 | 1725A | | Phyla cuneifolia (Torr.) Greene |
| 465 | 467 | 1725 | 1 | Phyla lanceolata (Michx.) Greene |
| 465 | 467 | 1724 | 1 | Verbena bracteata Lag. & Rodr. |
| 465 | 467 | 1721 | 1 | Verbena hastata L. |
| 465 | 467 | 1722 | 1 | Verbena simplex Lehm. |
| 465 | 467 | 1723 | 1 | Verbena stricta Vent. |
| 465 | 467 | 1720 | 1 | Verbena urticifolia L. var. leiocarpa Perry & Fern. |

VERBENACEAE     CONT.

465  467  1719  1 Verbena urticifolia L. var. urticifolia
465  467  1721A 1 Verbena X engelmannii Moldenke
465  467  1722A 1 Verbena X moechina Moldenke
465  467  1724A 1 Verbena X perriana Moldenke


VIOLACEAE

468  468  1432  1 Hybanthus concolor (T.F. Forst.) Spreng.
468  468  1439  1 Viola affinis Le Conte
468  468  1452  2 Viola arvensis Murr.
468  468  1450  1 Viola canadensis L.
468  468  1454  1 Viola conspersa Reichenb.
468  468  1442A 1 Viola fimbriatula Sm.
468  468  1441  1 Viola hirsutula Brainerd
468  468  1447  1 Viola incognita Brainerd
468  468  1443  1 Viola lanceolata L.
468  468  1446  1 Viola macloskeyi Lloyd ssp. pallens
                    (Banks ex DC.) M.S. Baker
468  468  1437  1 Viola missouriensis Greene
468  468  1437A 1 Viola missouriensis X sororia Brainerd
468  468  1437B 1 Viola missouriensis X triloba
468  468  1437C   Viola nephrophylla Greene
468  468  1436  1 Viola obliqua Hill
468  468  1433  1 Viola pedata L.
468  468  1434  1 Viola pedatifida G. Don
468  468  1444  1 Viola primulifolia L.
468  468  1448  1 Viola pubescens Ait. var. eriocarpa
                    (Schwein.) Russell
468  468  1449B   Viola pubescens Ait. var. leiocarpa
                    (Fern. & Wieg.) Seymour
468  468  1449A 1 Viola pubescens Ait. var. peckii House
468  468  1449  1 Viola pubescens Ait. var. pubescens
468  468  1451  2 Viola rafinesquii Greene
468  468  1455  1 Viola rostrata Pursh
468  468  1442  1 Viola sagittata Ait.
468  468  1438  1 Viola sororia Willd.
468  468  1453  1 Viola striata Ait.
468  468  1451A 2 Viola tricolor L.
468  468  1435  1 Viola triloba Schwein.
468  468  1435A 1 Viola triloba Schwein. var. dilatata
                    (Ell.) Brainerd
468  468  1434A 1 Viola X bernardii Greene
468  468  1436A 1 Viola X festata House
468  468  1442B 1 Viola X greenei House
468  468  1439A 1 Viola X slavinii House
468  468  1438B 1 Viola X variabilis Greene

## VITACEAE

```
471  471  1386  1 Ampelopsis cordata Michx.
471  471  1385  1 Parthenocissus inserta (Kern.)  Fritsch
471  471  1384  1 Parthenocissus quinquefolia (L.) Planch.
471  471  1384A   Parthenocissus tricuspidata (Sieb. &
                  Zucc.) Planch.
471  471  1379  1 Vitis aestivalis Michx.
471  471  1380  1 Vitis cinerea Engelm. ex Millard
471  471  1378  1 Vitis labrusca L.
471  471  1382  1 Vitis palmata Vahl
471  471  1383  1 Vitis riparia Michx.
471  471  1381  1 Vitis vulpina L.
```

## XYRIDACEAE

```
471  471  0587  1 Xyris caroliniana Walt.
471  471  0586  1 Xyris torta Sm.
```

## ZANNICHELLIACEAE

```
472  472  0098  1 Zannichellia palustris L.
```

## ZYGOPHYLLACEAE

```
473  473  1304A   Tribulus terrestris L.
```

# Taxon Names in Deam (1940) Which Have Undergone Nomenclatural Changes or Been Reduced to Synonymy or Which Deam Had Given the Incorrect Author

0003  1 BOTRYCHIUM MULTIFIDUM (GMEL.) RUPR. VAR. SILAIFOLIUM (PRESL) BROUN.  SEE:  Botrychium multifidum (Gmel.) Rupr. var. intermedium (D.C. Eat.) Farw.

0004A 1 BOTRYCHIUM DISSECTUM SPRENG. VAR. OBLIQUUM (MUHL.) CLUTE.  0004A  PROBABLY NOT DISTINCT

0014  1 DRYOPTERIS HEXAGONOPTERA (MICHX.) C. CHR.  SEE: Thelypteris hexagonoptera (Michx.) Weatherby

0015  1 DRYOPTERIS NOVEBORACENSIS (L.) A. GRAY.  SEE: Thelypteris noveboracensis (L.) Nieuwl.

0016  1 DRYOPTERIS THELYPTERIS (L.) A. GRAY VAR. PUBESCENS (LAWSON) A.R. PRINCE.  SEE:  Thelypteris palustris Schott var. pubescens (Lawson) Fern.

0021  1 DRYOPTERIS BOOTTII (TUCKERM.) UNDERW.  SEE:  Dryopteris X boottii (Tuckerman) Underwood

0024  1 ATHYRIUM PYCNOCARPON (SPRENG.) TIDESTR.  SEE:  Diplazium pycnocarpon (Spreng.) Broun

0025  1 ATHYRIUM THELYPTEROIDES (MICHX.) DESV.  SEE:  Deparia acrostichoides (Sw.) M. Kato ined.

0026  1 ATHYRIUM ASPLENIOIDES (MICHX.) DESV.  SEE:  Athyrium filix-femina (L.) Roth var. asplenioides (Michx.) Farw.

0027  1 ATHYRIUM ANGUSTUM (WILLD.) PRESL.  SEE:  Athyrium filix-femina (L.) Roth var. angustum (Willd.) Lawson

0027A 1 ATHYRIUM ANGUSTUM (WILLD.) PRESL VAR. ELATIUS (LINK) BUTTERS.  SEE:  Athyrium filix-femina (L.) Roth var. cyclosorum Rupr.

0028  1 CAMPTOSORUS RHIZOPHYLLUS (L.) LINK.  SEE:  Asplenium rhizophyllum L.

0031  1 ASPLENOSORUS EBENOIDES (SCOTT) WHERRY.  SEE:  Asplenium X ebenoides R.R. Scott

0033  1 ASPLENIUM CRYPTOLEPIS FERN.  SEE:  Asplenium ruta-muraria L. var. cryptolepis (Fern.) Wherry

0037  1 CHEILANTHES LANOSA (MICHX.) WATT.  SEE:  Cheilanthes lanosa (Michx.) D.C. Eat.

0039  1 PTERIDIUM LATIUSCULUM (DESV.) HIERON.  SEE:  Pteridium aquilinum (L.) Kuhn var. latiusculum (Desv.) Underwood ex Heller

73

0045  1 EQUISETUM VARIEGATUM SCHLEICH.   SEE:   Equisetum variegatum Schleich. ex Weber & C. Mohr.

0047  1 EQUISETUM PREALTUM RAF.   SEE:   Equisetum hyemale L. var. affine (Engelm.) A.A. Eat.

0049  1 EQUISETUM KANSANUM SCHAFFNER.          0049   PROBABLY NOT DISTINCT

0051  1 LYCOPODIUM SELAGO L. VAR. PATENS (BEAUV.) DESV.   SEE: Lycopodium selago L. var. selago

0055  1 LYCOPODIUM FLABELLIFORME (FERN.) BLANCHARD.   SEE: Lycopodium digitatum A. Braun

0067  1 JUNIPERUS COMMUNIS L. VAR. DEPRESSA PURSH.   SEE: Juniperus communis L.

0068  1 JUNIPERUS VIRGINIANA L. VAR. CREBRA FERNALD & GRISCOM. SEE:   Juniperus virginiana L.

0075  1 SPARGANIUM CHLOROCARPUM RYDB. VAR. ACAULE (BEEBY) FERN. 0075   PROBABLY NOT DISTINCT

0077  1 POTAMOGETON AMERICANUS CHAM. & SCHLECHT.          0077 PROBABLY NOT DISTINCT

0079  1 POTAMOGETON CAPILLACEUS POIR.   SEE:   Potamogeton diversifolius Raf.

0080  1 POTAMOGETON DIVERSIFOLIUS RAF.          0080   PROBABLY NOT DISTINCT

0081  1 POTAMOGETON GRAMINEUS L. VAR. GRAMINIFOLIUS FRIES.   SEE: Potamogeton gramineus L.

0083  1 POTAMOGETON ANGUSTIFOLIUS BERCH. AND PRESL.   SEE: Potamogeton illinoensis Morong

0086  1 POTAMOGETON LUCENS L.          0086   PROBABLY NOT DISTINCT

0092  1 POTAMOGETON FOLIOSUS RAF. VAR. GENUINUS.   SEE: Potamogeton foliosus Raf. var. foliosus

0092A 1 POTAMOGETON FOLIOSUS RAF. VAR. MACELLUS FERN. 0092A   PROBABLY NOT DISTINCT

0093  1 POTAMOGETON PUSILLUS L. VAR. MUCRONATUS (FIEBER) GRAEBN. SEE:   Potamogeton pusillus L. var. tenuissimus Mert. & Koch

0094  1 POTAMOGETON STRICTIFOLIUS BENNETT VAR. TYPICUS.    SEE:
      Potamogeton strictifolius Benn.

0095  1 POTAMOGETON PANORMITANUS BIV. VAR. MAJOR G. FISCHER.
      SEE:   Potamogeton pusillus L. var. pusillus

0095A 1 POTAMOGETON PANORMITANUS BIV. VAR. MINOR BIV.
      0095A  PROBABLY NOT DISTINCT

0098  1 ZANNICHELLIA PALUSTRIS L. VAR. MAJOR (BOENNINGH.) KOCH.
      SEE:   Zannichellia palustris L.

0101  1 NAJAS GRACILLIMA (A. BR.) MORONG.    SEE:    Najas
      gracillima (A. Braun) Magnus

0104  1 SCHEUCHZERIA PALUSTRIS L. VAR. AMERICANA FERN.    SEE:
      Scheuchzeria palustris L. ssp. americana (Fern.) Hulten

0105  1 ALISMA SUBCORDATUM RAF.    SEE:    Alisma plantago-aquatica
      L. var. parviflorum (Pursh) Torr.

0105A 1 ALISMA PLANTAGO-AQUATICA L. VAR. BREVIPES (GREENE)
      SAMUELSSON.    SEE:    Alisma plantago-aquatica L. var.
      americana Schultes & Schultes

0106  1 ECHINODORUS RADICANS (NUTT.) ENGELM.         0106
      PROBABLY NOT DISTINCT

0108  1 LOPHOTOCARPUS CALYCINUS (ENGELM.) J.G. SMITH.    SEE:
      Sagittaria calycina Engelm.

0116  1 ANACHARIS CANADENSIS (MICHX.) PLANCH.    SEE:    Elodea
      canadensis L.C. Rich.

0117  1 ANACHARIS OCCIDENTALIS (PURSH) VICT.    SEE:    Elodea
      nuttallii (Planch.) St. John

0119  1 ARUNDINARIA GIGANTEA (WALT.) CHAPM.    SEE:    Arundinaria
      gigantea (Walt.) Muhl.

0124  1 BROMUS LATIGLUMIS (SHEAR) HITCHC.    SEE:    Bromus
      altissimus Pursh

0125  1 BROMUS PURGANS L.         0125    PROBABLY NOT DISTINCT

0128  2 BROMUS MOLLIS L.    SEE:    Bromus hordeaceus L. ssp.
      hordeaceus

0131  1 FESTUCA OCTOFLORA WALT.    SEE:    Vulpia octoflora (Walt.)
      Rydb. var. octoflora

0132  1 FESTUCA OCTOFLORA WALT. VAR. TENELLA (WILLD.) FERN.    SEE:
      Vulpia octoflora (Walt.) Rydb. var. glauca (Nutt.) Fern.

0134  1 FESTUCA CAPILLATA LAM.    SEE:    Festuca tenuifolia
      Sibthorp

0137  1 FESTUCA OBTUSA SPRENG.    SEE:    Festuca obtusa Biehler

0142  1 GLYCERIA PALLIDA (TORR.) TRIN.    SEE:    Torreyochloa
      pallida (Torr.) Church

0148  1 POA AUTUMNALIS MUHL.    SEE:    Poa autumnalis Muhl. ex Ell.

0160  2 ERAGROSTIS CILIANENSIS (ALL.) LINK.    SEE:    Eragrostis
      cilianensis (All.) E. Mosher

0162  1 ERAGROSTIS FRANKII C.A. MEYER.    SEE:    Eragrostis frankii
      C.A. Mey. ex Steud.

0165  1 UNIOLA LATIFOLIA MICHX.    SEE:    Chasmanthium latifolium
      (Michx.) Yates

0167  1 PHRAGMITES COMMUNIS TRIN.    SEE:    Phragmites australis
      (Cav.) Trin. ex Steud.

0169  1 MELICA NITENS (SCRIBN.) NUTT.    SEE:    Melica nitens
      (Scribn.) Nutt. ex Piper

0171  1 TRIODIA FLAVA (L.) SMYTH.    SEE:    Tridens flavus (L.)
      A.S. Hitchc.

0175  1 AGROPYRON PAUCIFLORUM (SCHWEIN.) HITCHC.    SEE:
      Agropyron trachycaulum (Link) Malte ex H.F. Lewis var.
      trachycaulum

0176  1 AGROPYRON SUBSECUNDUM (LINK) HITCHC.    SEE:    Agropyron
      trachycaulum (Link) Malte ex H.F. Lewis var. unilaterale
      (Vasey) Malte

0179  1 ELYMUS VILLOSUS MUHL.    SEE:    Elymus villosus Muhl. ex
      Willd.

0180  1 ELYMUS VILLOSUS MUHL. FORMA ARKANSANUS (SCRIBN. & BALL)
      FERN.          0180    PROBABLY NOT DISTINCT

0182   1 ELYMUS VIRGINICUS L. VAR. INTERMEDIUS (VASEY) BUSH.
       0182   PROBABLY NOT DISTINCT

0185   1 ELYMUS VIRGINICUS L. VAR. JEJUNUS (RAMALEY) BUSH.
       0185   PROBABLY NOT DISTINCT

0186   1 ELYMUS VIRGINICUS L. VAR. GLABRIFLORUS (VASEY) BUSH.
       0186   PROBABLY NOT DISTINCT

0190   2 HORDEUM NODOSUM L.   SEE:   Hordeum brachyantherum Nevski

0195   1 SPHENOPHOLIS NITIDA (SPRENG.) SCRIBN.   SEE:
       Sphenopholis nitida (Biehler) Scribn.

0196   1 SPHENOPHOLIS INTERMEDIA (RYDB.) RYDB.   SEE:
       Sphenopholis obtusata (Michx.) Scribn. var. major (Torr.)
       K.S. Erdman

0197   1 SPHENOPHOLIS OBTUSATA (MICHX.) SCRIBN.          0197
       PROBABLY NOT DISTINCT

0199   2 ARRHENATHERUM ELATIUS (L.) MERT. & KOCH.   SEE:
       Arrhenatherum elatius (L.) Beauv. ex J. & C. Presl.

0201   1 DANTHONIA SPICATA (L.) BEAUV.   SEE:   Danthonia spicata
       (L.) Beauv. ex Roemer & Schultes

0203   1 CALAMAGROSTIS INEXPANSA GRAY.   SEE:   Calamagrostis
       neglecta (Ehrh.) Gaertn., Mey. & Scherb.

0206   2 AGROSTIS ALBA L.   SEE:   Agrostis stolonifera L. var.
       major (Gaudin) Farw.

0207   2 AGROSTIS PALUSTRIS HUDS.   SEE:   Agrostis stolonifera L.
       var. palustris (Huds.) Farw.

0220   1 MUHLENBERGIA CUSPIDATA (NUTT.) RYDB.   SEE:   Muhlenbergia
       cuspidata (Torr.) Rydb.

0222   1 MUHLENBERGIA MEXICANA (L.) TRIN.          0222   PROBABLY
       NOT DISTINCT

0223   1 MUHLENBERGIA MEXICANA (L.) TRIN. FORMA COMMUTATA (SCRIBN.)
       WIEG.   SEE:   Muhlenbergia mexicana (L.) Trin. f. commutata
       (Scribn.) Wieg.

0224   1 MUHLENBERGIA BRACHYPHYLLA BUSH.   SEE:   Muhlenbergia
       bushii Pohl

0227 1 MUHLENBERGIA FOLIOSA (ROEM. & SCHULT.) TRIN.   SEE:
Muhlenbergia mexicana (L.) Trin.

0227A 1 MUHLENBERGIA FOLIOSA (ROEM. & SCHULT.) TRIN. FORMA AMBIGUA
(TORR.) WIEG.   SEE:   Muhlenbergia frondosa (Poir.) Fern.

0228A 1 MUHLENBERGIA SYLVATICA TORR. FORMA ATTENUATA (SCRIBN.)
PALMER & STEYERMARK.            0228A  PROBABLY NOT DISTINCT

0229 1 SPOROBOLUS VAGINIFLORUS (TORR.) WOOD.   SEE:   Sporobolus
vaginiflorus (Torr. ex Gray) Wood

0230 1 SPOROBOLUS CLANDESTINUS (SPRENG.) HITCHC.   SEE:
Sporobolus clandestinus (Biehler) A.S. Hitchc.

0233 1 SPOROBOLUS HETEROLEPIS A. GRAY.   SEE:   Sporobolus
heterolepis (Gray) Gray

0237 1 ORYZOPSIS PUNGENS (TORR.) HITCHC.   SEE:   Oryzopsis
pungens (Torr. ex Spreng.) A.S. Hitchc.

0239 1 ORYZOPSIS RACEMOSA (J.E. SMITH) RICKER.   SEE:   Oryzopsis
racemosa (Sm.) Ricker

0246 1 ARISTIDA INTERMEDIA SCRIBN. & BALL.   SEE:   Aristida
longespica Poir. var. geniculata (Raf.) Fern.

0248 1 ARISTIDA RAMOSISSIMA ENGELM.   SEE:   Aristida ramosissima
Engelm. ex Gray

0251 1 LEPTOCHLOA PANICOIDES (PRESL) HITCHC.   SEE:   Diplachne
panicoides Presl (McNeill)

0259A 2 PHALARIS ARUNDINACEA L. VAR. PICTA L.   SEE:   Phalaris
arundinacea L.

0263A 1 ZIZANIA AQUATICA L. VAR. ANGUSTIFOLIA HITCHC.
0263A  PROBABLY NOT DISTINCT

0263B 1 ZIZANIA AQUATICA L. VAR. INTERIOR FASSETT.            0263B
PROBABLY NOT DISTINCT

0265 2 DIGITARIA ISCHAEMUM (SCHREB.) MUHL.   SEE:   Digitaria
ischaemum (Schreb. ex Schweig) Schreb. ex Muhl.

0267 1 LEPTOLOMA COGNATUM (SCHULT.) CHASE.   SEE:   Digitaria
cognatum (Schultes) Pilger

0269  1 PASPALUM CIRCULARE NASH.    SEE:    Paspalum laeve Michx.
var. circulare (Nash) Fern.

0270  1 PASPALUM PUBIFLORUM RUPR. VAR. GLABRUM VASEY.    SEE:
Paspalum pubiflorum Rupr. ex Fourn. var. glabrum Vasey Ex
Scribn.

0271  1 PASPALUM PUBESCENS MUHL.    SEE:    Paspalum setaceum Michx.
var. setaceum

0272  1 PASPALUM STRAMINEUM NASH.        0272    PROBABLY NOT
DISTINCT

0273  1 PANICUM DICHOTOMIFLORUM MICHX.    SEE:    Dichanthelium
dichotomum (L.) Gould var. dichotomum

0273A 1 PANICUM DICHOTOMIFLORUM MICHX. VAR. PURITANORUM SVENSON.
0273A  PROBABLY NOT DISTINCT

0281  1 PANICUM AGROSTOIDES SPRENG.    SEE:    Panicum rigidulum
Bosc ex Nees

0283  1 PANICUM DEPAUPERATUM MUHL.    SEE:    Dichanthelium
depauperatum (Muhl.) Gould

0283A 1 PANICUM DEPAUPERATUM MUHL. VAR. PSILOPHYLLUM FERN.
0283A  PROBABLY NOT DISTINCT

0284  1 PANICUM PERLONGUM NASH.    SEE:    Dichanthelium
linearifolium (Scribn.) Gould

0285  1 PANICUM LINEARIFOLIUM SCRIBN.        0285    PROBABLY NOT
DISTINCT

0286  1 PANICUM LINEARIFOLIUM SCRIBN. VAR. WERNER1 (SCRIBN.) FERN.
0286   PROBABLY NOT DISTINCT

0287  1 PANICUM XALAPENSE H.B.K.    SEE:    Dichanthelium laxiflorum
(Lam.) Gould

0288  1 PANICUM BICKNELLII NASH.    SEE:    Dichanthelium boreale
(Nash) Freckmann

0289  1 PANICUM MICROCARPON MUHL.    SEE:    Dichanthelium
sphaerocarpon (Ell.) Gould var. isophyllum (Scribn.) Gould &
Clark

0290  1 PANICUM DICHOTOMUM L.        0290    PROBABLY NOT
DISTINCT

0291  1 PANICUM MATTAMUSKEETENSE ASHE.        0291    PROBABLY
NOT DISTINCT

0292  1 PANICUM BOREALE NASH.         0292    PROBABLY NOT
DISTINCT

0293  1 PANICUM LUCIDUM ASHE.         0293    PROBABLY NOT
DISTINCT

0294  1 PANICUM YADKINENSE ASHE.          0294    PROBABLY NOT
DISTINCT

0295  1 PANICUM SPRETUM SCHULTES.    SEE:    Dichanthelium
acuminatum (Sw.) Gould var. densiflorum (Rand & Redf.) Gould
& Clar

0296  1 PANICUM LINDHEIMERI NASH.    SEE:    Dichanthelium
acuminatum (Sw.) Gould var. lindheimeri (Nash) Gould & Clark

0297  1 PANICUM AUBURNE ASHE.    SEE:    Dichanthelium acuminatum
(Sw.) Gould var. implicatum (Scribn.) Gould & Clark

0298  1 PANICUM PRAECOCIUS HITCHC. & CHASE.    SEE:    Dichanthelium
acuminatum (Sw.) Gould var. villosum (Gray) Gould & Clark

0299  1 PANICUM TENNESSEENSE ASHE.    SEE:    Dichanthelium
acuminatum (Sw.) Gould & Clark var. acuminatum (Sw.) Gould &
Clark

0300  1 PANICUM ALBEMARLENSE ASHE.        0300    PROBABLY NOT
DISTINCT

0301  1 PANICUM IMPLICATUM SCRIBN.        0301    PROBABLY NOT
DISTINCT

0302  1 PANICUM MERIDIONALE ASHE.        0302    PROBABLY NOT
DISTINCT

0303  1 PANICUM HUACHUCAE ASHE.         0303    PROBABLY NOT
DISTINCT

0304  1 PANICUM HUACHUCAE ASHE. VAR. FASCICULATUM (TORR.) F.T.
HUBB.          0304    PROBABLY NOT DISTINCT

0305  1 PANICUM SUBVILLOSUM ASHE.        0305    PROBABLY NOT
DISTINCT

0306  1 PANICUM SCOPARIOIDES ASHE.        0306    PROBABLY NOT
DISTINCT

0307  1 PANICUM VILLOSISSIMUM NASH.        0307    PROBABLY NOT
DISTINCT

0308  1 PANICUM PSEUDOPUBESCENS NASH.       0308    PROBABLY NOT
DISTINCT

0309  1 PANICUM DEAMII HITCH. & CHASE.    SEE:    Dichanthelium
consanguineum (Kunth) Gould & Clark

0310  1 PANICUM ADDISONII NASH.    SEE:    Dichanthelium ovale
(Ell.) Gould & Clark var. ovale

0311  1 PANICUM TSUGETORUM NASH.    SEE:    Dichanthelium sabulorum
(Lam.) Gould & Clark var. patulum (Scribn. & Merr.) Gould

0312  1 PANICUM COLUMBIANUM SCRIBN.    SEE:    Dichanthelium
sabulorum (Lam.) Gould & Clark var. thinium (A.S. Hitchc. &
Chase)

0313  1 PANICUM POLYANTHES SCHULTES.        0313    PROBABLY NOT
DISTINCT

0314  1 PANICUM SPHAEROCARPON ELL.    SEE:    Dichanthelium
sphaerocarpon (Ell.) Gould var. sphaerocarpon

0315  1 PANICUM LEIBERGII (VASEY) SCRIBN.    SEE:    Dichanthelium
leibergii (Vasey) Freckmann

0316  1 PANICUM OLIGOSANTHES SCHULTES.    SEE:    Dichanthelium
oligosanthes (Schultes) Gould var. oligosanthes

0317  1 PANICUM SCRIBNERIANUM NASH.    SEE:    Dichanthelium
oligosanthes (Schultes) Gould var. wilcoxianum (Vasey) Gould
& Clark

0318  1 PANICUM ASHEI PEARSON.        0318    PROBABLY NOT
DISTINCT

0319  1 PANICUM COMMUTATUM SCHULTES.    SEE:    Dichanthelium
commutatum (Schultes) Gould

0320  1 PANICUM CLANDESTINUM L.    SEE:    Dichanthelium
clandestinum (L.) Gould

0321  1 PANICUM LATIFOLIUM L.    SEE:    Dichanthelium latifolium
(L.) Gould & Clark

0322  1 PANICUM BOSCII POIR.    SEE:    Dichanthelium boscii (Poir.)
Gould & Clark

0323  1 PANICUM BOSCII POIR. VAR. MOLLE (VASEY) HITCHC. & CHASE.
      0323   PROBABLY NOT DISTINCT

0325A 1 ECHINOCHLOA WALTERI (PURSH) HELLER FORMA LAEVIGATA WIEG.
      0325A  PROBABLY NOT DISTINCT

0326  2 SETARIA LUTESCENS (WEIGEL) F.T. HUBB.   SEE:   Setaria
      glauca (L.) Beauv.

0330  1 CENCHRUS PAUCIFLORUS BENTH.   SEE:   Cenchrus incertus
      M.A. Curtis

0331  1 ANDROPOGON SCOPARIUS MICHX.   SEE:   Schizachyrium
      scoparium (Michx.) Nash

0332  1 ANDROPOGON FURCATUS MUHL.   SEE:   Andropogon gerardii
      Vitman var. gerardii

0337  1 TRIPSACUM DACTYLOIDES L.   SEE:   Tripsacum dactyloides
      (L.) L.

0338  1 HEMICARPHA MICRANTHA (VAHL) PAX.   SEE:   Hemicarpha
      micrantha (Vahl) Britt.

0344  1 CYPERUS INFLEXUS MUHL.   SEE:   Cyperus aristatus Rottb.
      var. aristatus

0354  1 CYPERUS STRIGOSUS L.        0354   PROBABLY NOT DISTINCT

0356  1 CYPERUS FERRUGINESCENS BOECKL.   SEE:   Cyperus odoratus
      L. var. odoratus

0358  1 KYLLINGA PUMILA MICHX.   SEE:   Cyperus tenuifolius
      (Steud.) Dandy

0359  1 ERIOPHORUM SPISSUM FERN.   SEE:   Eriophorum vaginatum L.
      ssp. spissum (Fern.) Hulten

0361  1 ERIOPHORUM ANGUSTIFOLIUM ROTH.   SEE:   Eriophorum
      angustifolium Honckeny

0364  1 FUIRENA PUMILA TORR.   SEE:   Fuirena pumila (Torr.)
      Spreng.

0366  1 SCIRPUS DEBILIS PURSH.   SEE:   Scirpus purshianus Fern.

0368  1 SCIRPUS SMITHII GRAY VAR. SETOSUS FERN.          0368
      PROBABLY NOT DISTINCT

0371  1 SCIRPUS VALIDUS VAHL.     SEE:   Scirpus tabernaemontanii
      K.C. Gmel.

0374  1 SCIRPUS ATROVIRENS MUHL.     SEE:   Scirpus atrovirens
      Willd.

0375  1 SCIRPUS ATROVIRENS MUHL. VAR. GEORGIANUS (HARPER) FERN.
      SEE:   Scirpus georgianus Harper

0381  1 ELEOCHARIS QUADRANGULATA (MICHX.) R. & S. VAR. CRASSIOR
      FERN.   SEE:   Eleocharis quadrangulata (Michx.) Roemer &
      Schultes

0385  1 ELEOCHARIS OVATA (ROTH) R. & S.   SEE:   Eleocharis obtusa
      (Willd.) Schultes var. ovata (Roth) Drapalik & Mohlenbrock

0386  1 ELEOCHARIS INTERMEDIA (MUHL.) SCHULTES.   SEE:
      Eleocharis intermedia Schultes

0390  1 ELEOCHARIS CALVA TORR.     SEE:   Eleocharis erythropoda
      Steud.

0391  1 ELEOCHARIS ACICULARIS (L.) R. & S. VAR. TYPICA SVENSON.
      SEE:   Eleocharis acicularis (L.) Roemer & Schultes var.
      acicularis

0394  1 ELEOCHARIS MICROCARPA TORR. VAR. FILICULMIS TORR.   SEE:
      Eleocharis microcarpa Torr.

0396  1 ELEOCHARIS TENUIS (WILLD.) SCHULTES VAR. VERRUCOSA (SVEN.)
      SVEN.   SEE:   Eleocharis verrucosa (Svens.) L. Harms

0397  1 ELEOCHARIS COMPRESSA SULLIV.          0397    PROBABLY NOT
      DISTINCT

0398  1 ELEOCHARIS ROSTELLATA TORR.   SEE:   Eleocharis rostellata
      (Torr.) Torr.

0399  1 ELEOCHARIS PAUCIFLORA (LIGHTF.) LINK VAR. FERNALDII SVEN.
      SEE:   Eleocharis pauciflora (Lightf.) Link

0400  1 FIMBRISTYLIS PUBERULA (MICHX.) VAHL.   SEE:   Fimbristylis
      caroliniana (Lam.) Fern.

0401  1 FIMBRISTYLIS AUTUMNALIS (L.) R. & S. VAR. MUCRONULATA
      (MICHX.) FERN.   SEE:   Fimbristylis autumnalis (L.) Roemer &
      Schultes

0402A 1 BULBOSTYLIS CAPILLARIS (L.) BRITT. VAR. CREBRA FERN.
0402A   PROBABLY NOT DISTINCT

0408  1 RHYNCHOSPORA CAPILLACEA TORR. FORMA. LEVISETA (E.J. HILL)
FERN.          0408   PROBABLY NOT DISTINCT

0410  1 RHYNCHOSPORA CYMOSA ELL.   SEE:   Rhynchospora globularis
(Chapm.) Small var. globularis

0411  1 RHYNCHOSPORA GLOMERATA (L.) VAHL VAR. MINOR BRITT.   SEE:
Rhynchospora capitellata (Michx.) Vahl

0414  1 SCLERIA RETICULARIS MICHX.   SEE:   Scleria reticularis
Michx. var. reticularis

0415  1 SCLERIA SETACEA POIR.          0415   PROBABLY NOT
DISTINCT

0417  1 SCLERIA VERTICILLATA MUHL.   SEE:   Scleria verticillata
Muhl. ex Willd.

0418  1 CAREX SARTWELLII DEWEY VAR. STENORRHYNCHA HERMANN.   SEE:
Carex sartwellii Dewey

0419  1 CAREX SICCATA DEWEY.   SEE:   Carex foenea Willd. var.
foenea

0420  1 CAREX CHORDORRHIZA L.F.   SEE:   Carex chordorrhiza Ehrh.
ex L. f.

0421  1 CAREX RETROFLEXA MUHL.   SEE:   Carex retroflexa Willd.

0422  1 CAREX ROSEA SCHKUHR.   SEE:   Carex rosea Willd.

0424  1 CAREX CEPHALOPHORA MUHL.   SEE:   Carex cephalophora
Willd.

0427  1 CAREX MUHLENBERGII SCHKUHR. VAR. ENERVIS.   SEE:   Carex
muhlenbergii Willd. var. muhlenbergii

0431  1 CAREX CEPHALOIDEA DEWEY.   SEE:   Carex cephaloidea
(Dewey) Dewey

0433  1 CAREX SPARGANIOIDES MUHL.   SEE:   Carex sparganioides
Willd.

0434  1 CAREX ANNECTENS BICKN.    SEE:    Carex annectens (Bickn.)
Bickn.

0439  1 CAREX STIPATA MUHL.    SEE:    Carex stipata Muhl. ex Willd.
var. stipata

0442  1 CAREX CRUS-CORVI SHUTTLW.    SEE:    Carex crus-corvi
Schuttlw. ex Kunze

0447  1 CAREX CANESCENS L. VAR. DISJUNCTA FERN.    SEE:    Carex
canescens L. ssp. arctiformis (Mackenzie) Calder & Taylor
var. disjuncta Fern.

0448  1 CAREX CANESCENS L. VAR. SUBLOLIACEA LAEST.    SEE:    Carex
canescens L. ssp. canescens var. subloliacea (Laestad.)
Hartman

0452  1 CAREX INCOMPERTA BICKN.    SEE:    Carex atlantica Bailey
var. incomperta (Bickn.) F.J. Herm.

0454  1 CAREX LARICINA MACK.    SEE:    Carex cephalantha (Bailey)
Bickn.

0455  1 CAREX BROMOIDES SCHKUHR.    SEE:    Carex bromoides Willd.

0457  1 CAREX BEBBII OLNEY.    SEE:    Carex bebbii (Bailey) Fern.

0460  1 CAREX FESTUCACEA SCHKUHR.    SEE:    Carex festucacea Willd.

0462  1 CAREX BREVIOR (DEWEY) MACK.    SEE:    Carex brevior (Dewey)
Mackenzie ex Lunell

0465  1 CAREX RICHII (FERN.) MACK.    SEE:    Carex straminea Willd.

0466  1 CAREX CUMULATA (BAILEY) MACK.    SEE:    Carex cumulata
(Bailey) Fern.

0473  1 CAREX LEPTALEA WAHL.    SEE:    Carex leptalea Wahlenb. var.
leptalea

0474  1 CAREX LEPTALEA WAHL. VAR. HARPERI (FERN.) STONE.    SEE:
Carex leptalea Wahlenb. ssp. harperi (Fern.) Calder & Taylor

0477  1 CAREX ARTITECTA MACK.    SEE:    Carex artitecta Mackenzie
var. artitecta

0484  1 CAREX UMBELLATA SCHKUHR.    SEE:    Carex umbellata Schkuhr
ex Willd.

0497  1 CAREX CAREYANA TORR.    SEE:    Carex careyana Dewey

0509  1 CAREX HALEANA OLNEY.    SEE:    Carex granularis Muhl. ex
Willd. var. haleana (Olney) Porter

0512  1 CAREX OLIGOCARPA SCHKUHR.    SEE:    Carex oligocarpa Willd.

0514  1 CAREX CONOIDEA SCHKUHR.    SEE:    Carex conoidea Willd.

0516  1 CAREX GRISEA WAHL.    SEE:    Carex amphibola Steud. var.
turgida Fern.

0517  1 CAREX GLAUCODEA TUCKERM.    SEE:    Carex flaccosperma Dewey
var. glaucodea (Tuckerman) Kukenth.

0523  1 CAREX SPRENGELII DEWEY.    SEE:    Carex sprengelii Dewey ex
Spreng.

0529  1 CAREX VIRESCENS MUHL.        0529    PROBABLY NOT
DISTINCT

0538  1 CAREX SUBSTRICTA (KUKENTH.) MACK.    SEE:    Carex aquatilis
Wahlenb. var. aquatilis

0546  1 CAREX HYSTRICINA MUHL.    SEE:    Carex hystricina Muhl. ex
Willd.

0549  1 CAREX RIPARIA CURTIS CURTIS VAR. LACUSTRIS (WILLD.)
KUKENTH.    SEE:    Carex lacustris Willd.

0553  1 CAREX TRICHOCARPA MUHL.    SEE:    Carex trichocarpa Schkuhr

0558  1 CAREX ROSTRATA STOKES.    SEE:    Carex rostrata Stokes ex
With.

0559  1 CAREX TUCKERMANI BOOTT.    SEE:    Carex tuckermanii Dewey

0565  1 CAREX INTUMESCENS RUDGE VAR. FERNALDII BAILEY.
0565    PROBABLY NOT DISTINCT

0567  1 CAREX LUPULINA MUHL.    SEE:    Carex lupulina Willd.

0568  1 CAREX LUPULIFORMIS SARTWELL.    SEE:    Carex lupuliformis Sartwell ex Dewey

0570  1 ACORUS CALAMUS L.    SEE:    Acorus americanus (Raf.) Raf.

0573  1 PELTANDRA VIRGINICA (L.) KUNTH.    SEE:    Peltandra virginica (L.) Schott ssp. virginica

0575  1 ARISAEMA PUSILLUM (PECK) NASH.    SEE:    Arisaema triphyllum (L.) Schott ssp. pusillum (Peck) Huttleston

0576  1 ARISAEMA TRIPHYLLUM (L.) SCHOTT.    SEE:    Arisaema triphyllum (L.) Schott ssp. triphyllum

0578  1 LEMNA TRISULCA L.         0578    PROBABLY NOT DISTINCT

0579  1 LEMNA MINIMA PHILLIPI.    SEE:    Lemna minuta H.B.K.

0582  1 LEMNA CYCLOSTASA (ELL.) CHEVALIER.    SEE:    Lemna valdiviana Phil.

0584  1 WOLFFIA PUNCTATA GRISEB.    SEE:    Wolffia borealis (Engelm.) Landolt

0585  1 WOLFFIELLA FLORIDANA (J.D. SMITH) THOMPSON.    SEE:    Wolffiella gladiata (Hegelm.) Hegelm.

0592  1 COMMELINA ANGUSTIFOLIA MICHX.    SEE:    Commelina erecta L. var. deamiana Fern.

0593  1 TRADESCANTIA CANALICULATA RAF.    SEE:    Tradescantia ohiensis Raf. var. ohiensis

0594  1 TRADESCANTIA SUBASPERA KER VAR. TYPICA ANDERSON & WOODSON. SEE:    Tradescantia subaspera Ker-Gawl. var. subaspera

0604  1 JUNCUS SECUNDUS BEAUV.    SEE:    Juncus secundus Beauv. ex Poir.

0605  1 JUNCUS MACER S.F. GRAY.    SEE:    Juncus tenuis Willd. var. tenuis

0605A 1 JUNCUS MACER S.F. GRAY FORMA WILLIAMSII (FERN.) HERMANN.
SEE:   Juncus tenuis Willd. var. williamsii Fern.

0605B 1 JUNCUS MACER S.F. GRAY FORMA ANTHELATUS (WIEG.) HERMANN.
0605B   PROBABLY NOT DISTINCT

0605C 1 JUNCUS MACER S.F. GRAY FORMA DISCRETIFLORUS HERMANN.
0605C   PROBABLY NOT DISTINCT

0607   1 JUNCUS DUDLEYI WIEG.     SEE:     Juncus tenuis Willd. var.
uniflorus (Farw.) Farw.

0609   1 JUNCUS BIFLORUS ELL.        0609    PROBABLY NOT DISTINCT

0610   1 JUNCUS CANADENSIS J. GAY.     SEE:    Juncus canadensis J.
Gay ex Laharpe

0621   1 JUNCUS ALPINUS VILL. VAR. RARIFLORUS HARTM.    SEE:
Juncus alpinus Vill. ssp. nodulosus (Wahlenb.) Lindm. var.
rariflorus Hartman

0621A 1 JUNCUS ALPINUS VILL. VAR. FUSCESCENS FERN.    SEE:    Juncus
alpinus Vill. ssp. alpinus

0622   1 LUZULA CAROLINAE S. WATS. VAR. SALTUENSIS (FERN.) FERN.
SEE:   Luzula acuminata Raf. var. acuminata

0623   1 LUZULA MULTIFLORA (EHRH.) LEJEUNE.    SEE:    Luzula
multiflora (Retz.) Lej.

0628   1 STENANTHIUM GRAMINEUM (KER) KUNTH.    SEE:    Stenanthium
gramineum (Ker-Gawl.) Morong

0629   1 STENANTHIUM ROBUSTUM WATS.    SEE:    Stenanthium gramineum
(Ker-Gawl.) Morong var. robustum (S. Wats.) Fern.

0630   1 ZIGADENUS GLAUCUS NUTT.    SEE:    Zigadenus elegans Pursh
ssp. glaucus (Nutt.) Hulten

0635   2 HEMEROCALLIS FULVA L.    SEE:    Hemerocallis fulva (L.) L.

0639   1 ALLIUM CANADENSE L.        0639    PROBABLY NOT DISTINCT

0647   1 ERYTHRONIUM AMERICANUM KER.    SEE:    Erythronium
americanum Ker-Gawl.

0652  1 SMILACINA RACEMOSA (L.) DESF. VAR. TYPICA FERN.   SEE:
Smilacina racemosa (L.) Desf.

0653  1 SMILACINA STELLATA (L.) DESF.      0653   PROBABLY NOT
DISTINCT

0654  1 MAIANTHEMUM CANADENSE DESF.      0654   PROBABLY NOT
DISTINCT

0657A 1 POLYGONATUM CANALICULATUM (MUHL.) PURSH.        0657A
PROBABLY NOT DISTINCT

0664A 1 TRILLIUM GLEASONI FERN. FORMA WALPOLEI (FARW.) DEAM.
0664A  PROBABLY NOT DISTINCT

0668  1 SMILAX HERBACEA L. VAR. LASIONEURA (HOOK.) A. DC.   SEE:
Smilax lasioneuron Hook.

0669  1 SMILAX ECIRRHATA (ENGELM.) WATS.   SEE:   Smilax ecirrhata
(Engelm. ex Kunth) S. Wats.

0670  1 SMILAX GLAUCA WALT. VAR. GENUINA BLAKE.   SEE:   Smilax
glauca Walt. var. glauca

0674  1 HYMENOCALLIS OCCIDENTALIS (LE CONTE) KUNTH.   SEE:
Hymenocallis caroliniana (L.) Herbert

0675  1 AGAVE VIRGINICA L.   SEE:   Manfreda virginica (L.) Rose

0679  1 DIOSCOREA GLAUCA MUHL.   SEE:   Dioscorea quaternata
(Walt.) J.F. Gmel.

0680  1 DIOSCOREA QUATERNATA (WALT.) GMEL.      0680
PROBABLY NOT DISTINCT

0681  1 IRIS CRISTATA AIT.   SEE:   Iris cristata Soland.

0686  1 SISYRINCHIUM GRAMINOIDES BICKN.   SEE:   Sisyrinchium
angustifolium P. Mill.

0689  1 CYPRIPEDIUM CANDIDUM MUHL.   SEE:   Cypripedium candidum
Muhl. ex Willd.

0691  1 CYPRIPEDIUM PARVIFLORUM SALISB. VAR. PUBESCENS (WILLD.)
KNIGHT.   SEE:   Cypripedium pubescens Willd.

0693  1 ORCHIS SPECTABILIS L.    SEE:    Galearis spectabilis (L.)
      Raf.

0694  1 HABENARIA VIRIDIS (L.) R. BR. VAR. BRACTEATA (MUHL.) GRAY.
      SEE:   Coeloglossum viride (L.) Hartman

0695  1 HABENARIA FLAVA (L.) GRAY.    SEE:   Platanthera flava (L.)
      Lindl. var. flava

0696  1 HABENARIA SCUTELLATA (NUTT.) F. MORRIS.         0696
      PROBABLY NOT DISTINCT

0697  1 HABENARIA DILATATA (PURSH) GRAY.    SEE:    Platanthera
      dilatata (Pursh) Lindl. ex Beck

0698  1 HABENARIA HYPERBOREA (L.) R. BR.    SEE:    Platanthera
      hyperborea (L.) Lindl.

0699  1 HABENARIA CLAVELLATA (MICHX.) SPRENG.    SEE:    Platanthera
      clavellata (Michx.) Luer

0700  1 HABENARIA ORBICULATA (PURSH) TORR.    SEE:    Platanthera
      orbiculata (Pursh) Lindl.

0701  1 HABENARIA HOOKERI TORR.    SEE:    Platanthera hookeri
      (Torr. ex Gray) Lindl.

0702  1 HABENARIA CILIARIS (L.) R. BR.    SEE:    Platanthera
      ciliaris (L.) Lindl.

0703  1 HABENARIA LACERA (MICHX.) LODD.    SEE:    Platanthera
      lacera (Michx.) G. Don

0704  1 HABERNARIA LEUCOPHAEA (NUTT.) GRAY.    SEE:    Platanthera
      leucophaea (Nutt.) Lindl.

0705  1 HABENARIA PSYCODES (L.) SPRENG.    SEE:    Platanthera
      psycodes (L.) Lindl.

0706  1 HABENARIA PERAMOENA GRAY.    SEE:    Platanthera peramoena
      (Gray) Gray

0707  1 POGONIA OPHIOGLOSSOIDES (L.) KER.    SEE:    Pogonia
      ophioglossoides (L.) Juss.

0709  1 ISOTRIA VERTICILLATA (WILLD.) RAF.    SEE:    Isotria
      verticillata (Muhl. ex Willd.) Raf.

0711  1 EPIPACTIS LATIFOLIA (HUDS.) ALL.   SEE:   Epipactis
helleborine (L.) Crantz

0712  1 SPIRANTHES BECKII LINDL.   SEE:   Spiranthes lacera (Raf.)
Raf. var. gracilis (Bigelow) Luer

0713  1 SPIRANTHES GRACILIS (BIGEL.) BECK.         0713
PROBABLY NOT DISTINCT

0717  1 GOODYERA PUBESCENS R. BR.   SEE:   Goodyera pubescens
(Willd.) R. Br.

0718  1 CALOPOGON PULCHELLUS (SALISB.) R. BR.   SEE:   Calopogon
tuberosus (L.) B.S.P.

0720  1 CORALLORRHIZA MACULATA RAF.   SEE:   Corallorhiza maculata
(Raf.) Raf.

0721  1 CORALLORRHIZA ODONTORHIZA NUTT.   SEE:   Corallorhiza
odontorhiza (Willd.) Nutt.

0723  1 LIPARIS LILIIFOLIA (L.) RICHARD.   SEE:   Liparis
lilifolia (L.) L. C. Rich. ex Lindl.

0727  1 APLECTRUM HYEMALE (MUHL.) TORR.   SEE:   Aplectrum hyemale
(Muhl. ex Willd.) Nutt.

0730  1 POPULUS DELTOIDES MICHX.   SEE:   Populus deltoides Bartr.
ex Marsh.

0737  1 SALIX LUCIDA MUHL. VAR. INTONSA FERN.         0737
PROBABLY NOT DISTINCT

0738  1 SALIX LONGIPES SHUTTLW. VAR. WARDI (BEBB) SCHNEID.   SEE:
Salix caroliniana Michx.

0741  1 SALIX INTERIOR ROWLEE.         0741   PROBABLY NOT
DISTINCT

0742  1 SALIX INTERIOR ROWLEE. VAR. WHEELERI ROWLEE.   SEE:
Salix exigua Nutt.

0743  1 SALIX DISCOLOR MUHL.         0743   PROBABLY NOT DISTINCT

0744  1 SALIX DISCOLOR MUHL. VAR. LATIFOLIA ANDERS.   SEE:   Salix
discolor Muhl.

0748  1 SALIX TRISTIS AIT.   SEE:   Salix humilis Marsh. var. microphylla (Anderss.) Fern.

0750  1 SALIX PEDICELLARIS PURSH VAR. HYPOGLAUCA FERN.   SEE: Salix pedicellaris Pursh

0751  1 SALIX CANDIDA FLUGGE.   SEE:   Salix candida Flugge ex Willd.

0752  1 SALIX ADENOPHYLLA HOOK.   SEE:   Salix cordata Muhl.

0753  1 SALIX CORDATA MUHL.        0753   PROBABLY NOT DISTINCT

0754  1 SALIX GLAUCOPHYLLA BEBB.   SEE:   Salix myricoides Muhl. var. myricoides

0758  1 CARYA PECAN (MARSH.) ENGLER & GRAEBNER.   SEE:   Carya illinoensis (Wang.) K. Koch

0762  1 CARYA TOMENTOSA (LAM.) NUTT.   SEE:   Carya tomentosa (Poir.) Nutt.

0766  1 CARYA BUCKLEYI DURAND VAR. ARKANSANA SARG.   SEE:   Carya texana Buckl.

0767  1 CARPINUS CAROLINIANA WALT. VAR. VIRGINIANA (MARSH.) FERN. SEE:   Carpinus caroliniana Walt.

0769  1 OSTRYA VIRGINIANA L. FORMA GLANDULOSA (SPACH) MAC BR. SEE:   Ostrya virginiana (P. Mill.) K. Koch f. glandulosa (Spach) Macbr.

0776  1 ALNUS INCANA (L.) MOENCH VAR. AMERICANA REGEL. 0776   PROBABLY NOT DISTINCT

0777  1 ALNUS RUGOSA (EHRH.) SPRENG.   SEE:   Alnus incana (L.) Moench ssp. rugosa (Du Roi) Clausen

0784  1 QUERCUS PRINUS L.   SEE:   Quercus montana Willd.

0790  1 QUERCUS BOREALIS MICHX. VAR. MAXIMA (MARSH.) ASHE.   SEE: Quercus rubra L. var. rubra

0798  1 ULMUS FULVA MICHX.   SEE:   Ulmus rubra Muhl.

0799  1 ULMUS RACEMOSA THOMAS.   SEE:   Ulmus thomasii Sarg.

0803  1 CELTIS OCCIDENTALIS L. VAR. CRASSIFOLIA (LAM.) GRAY.
SEE:   Celtis occidentalis L. var. occidentalis

0805  1 CELTIS PUMILA (MUHL.) PURSH.   SEE:   Celtis occidentalis
L. var. pumila (Pursh) Gray

0807  1 MORUS ALBA L. VAR. TATARICA (L.) LOUD.   SEE:   Morus alba
L.

0810  1 HUMULUS AMERICANUS NUTT.   SEE:   Humulus lupulus L. var.
lupuloides E. Small

0811  1 CANNABIS SATIVA L.   SEE:   Cannabis sativa L. ssp. sativa
var. sativa

0812  1 URTICA PROCERA MUHL.   SEE:   Urtica dioica L. ssp.
gracilis (Ait.) Seland. var. procera (Muhl.) Weddell

0813  1 LAPORTEA CANADENSIS (L.) GAUD.   SEE:   Laportea
canadensis (L.) Weddell

0816  1 BOEHMERIA CYLINDRICA (L.) SW.   SEE:   Boehmeria
cylindrica (L.) Sw. var. cylindrica

0817  1 BOEHMERIA CYLINDRICA (L.) SW. VAR. DRUMMONDIANA WEDDELL.
SEE:   Boehmeria cylindrica (L.) Sw. var. drummondiana
(Weddell) Weddell

0818  1 PARIETARIA PENNSYLVANICA MUHL.   SEE:   Parietaria
pensylvanica Muhl. ex Willd.

0819  1 PHORADENDRON FLAVESCENS (PURSH) NUTT.   SEE:
Phoradendron serotinum (Raf.) M.C. Johnston

0820  1 COMANDRA RICHARDSIANA FERN.   SEE:   Comandra umbellata
(L.) Nutt. ssp. umbellata

0821  1 ASARUM REFLEXUM BICKN.   SEE:   Asarum canadense L. var.
reflexum (Bickn.) B.L. Robins.

0829  1 RUMEX BRITANNICA L.   SEE:   Rumex orbiculatus Gray

0832  1 POLYGONUM EXSERTUM SMALL.   SEE:   Polygonum ramosissimum
Michx. var. ramosissimum.

0834  1 POLYGONUM MONSPELIENSE THIEBAUD.    SEE:    Polygonum
aviculare L. var. vegetum Ledeb.

0835  1 POLYGONUM BUXIFORME SMALL.         0835    PROBABLY NOT
DISTINCT

0836  1 POLYGONUM AVICULARE L.            0836    PROBABLY NOT
DISTINCT

0837  1 POLYGONUM NEGLECTUM BESSER.        0837    PROBABLY NOT
DISTINCT

0838  1 POLYGONUM TENUE MICHX.            0838    PROBABLY NOT
DISTINCT

0839  1 POLYGONUM NATANS A. EATON FORMA GENUINUM STANFORD.    SEE:
Polygonum amphibium L. var. stipulaceum (Coleman) Fern.

0840  1 POLYGONUM NATANS A. EATON FORMA HARTWRIGHTII (GRAY)
STANFORD.          0840    PROBABLY NOT DISTINCT

0841  1 POLYGONUM COCCINEUM MUHL.    SEE:    Polygonum amphibium L.
var. emersum Michx.

0842  1 POLYGONUM PENNSYLVANICUM L. VAR. GENUINUM FERN.    SEE:
Polygonum pensylvanicum (L.) Small

0844  1 POLYGONUM PENNSYLVANICUM L. VAR. LAEVIGATUM FORMA
PALLESCENS STANFORD.    0844    FORMA NOT DISTINCT

0847  1 POLYGONUM HYDROPIPER L. VAR. PROJECTUM STANFORD.    SEE:
Polygonum hydropiper L.

0853  1 POLYGONUM ARIFOLIUM L. VAR. LENTIFORME FERNALD & GRISCOM.
SEE:    Polygonum arifolium L. var. pubescens (Keller) Fern.

0856  1 POLYGONUM DUMETORUM L.    SEE:    Polygonum scandens L. var.
dumetorum (L.) Gleason

0860  2 CHENOPODIUM AMBROSIOIDES L. SSP. EU-AMBROSIOIDES AELLEN.
SEE:    Chenopodium ambrosioides L.

0861  2 CHENOPODIUM AMBROSIOIDES L. SSP. EU-AMBROSIOIDES VAR.
ANTHELMINTICUM    (L.) AELLEN.          0861    PROBABLY NOT
DISTINCT

0863  2 CHENOPODIUM GLAUCUM L. SSP. EU-GLAUCUM AELLEN.    SEE:
Chenopodium glaucum L. var. glaucum

0867 1 CHENOPODIUM PRATERICOLA RYDB. SEE: Chenopodium desiccatum A. Nels. leptophylloides (J. Murr) H.A. Wahl

0874A 1 ATRIPLEX PATULA L. VAR. HASTATA (L.) GRAY. 0874A PROBABLY NOT DISTINCT

0875 1 ATRIPLEX PATULA L. VAR. LITTORALIS (L.) A. GRAY. SEE: Atriplex littoralis L.

0876 1 CORISPERMUM NITIDUM KIT. SEE: Corispermum nitidum Kit. ex Schultes

0878 2 SALSOLA PESTIFER A. NELSON. SEE: Salsola kali L.

0883 1 AMARANTHUS GRAECIZANS L. SEE: Amaranthus albus L.

0884 1 ACNIDA TAMARISCINA (NUTT.) WOOD. SEE: Amaranthus rudis Sauer

0885 1 ACNIDA ALTISSIMA RIDDELL. SEE: Amaranthus tuberculatus (Moq.) Sauer

0886 1 ACNIDA SUBNUDA (S. WATS.) STANDLEY. 0886 PROBABLY NOT DISTINCT

0887A 2 FROELICHIA CAMPESTRIS SMALL. SEE: Froelichia floridana (Nutt.) Moq. var. campestris (Small) Fern.

0889 1 OXYBAPHUS NYCTAGINEUS (MICHX.) SWEET. SEE: Mirabilis nyctaginea (Michx.) MacM.

0896 1 STELLARIA LONGIFOLIA MUHL. SEE: Stellaria longifolia Muhl ex Willd.

0898 1 STELLARIA PUBERA MICHX. VAR. SILVATICA (BEGUINOT) WEATHERBY. SEE: Stellaria corei Shinners

0900 2 CERASTIUM VULGATUM L. VAR. HIRSUTUM FRIES. SEE: Cerastium fontanum Baumg. ssp. triviale (Link) Jalas

0901 2 CERASTIUM VULGATUM L. VAR. HIRSUTUM FORMA GLANDULOSUM (BOENN.) DRUCE. 0901 PROBABLY NOT DISTINCT

0903 2 CERASTIUM VISCOSUM L. SEE: Cerastium glomeratum Thuill.

0907  1 ARENARIA LATERIFLORA L.    SEE:    Moehringia lateriflora
      (L.) Fenzl

0908  1 ARENARIA STRICTA MICHX.    SEE:    Minuartia stricta (Sw.)
      Hiern

0909  1 ARENARIA PATULA MICHX.    SEE:    Minuartia patula (Michx.)
      Mattf.

0911  1 PARONYCHIA FASTIGIATA (RAF.) FERN. VAR. TYPICA FERN.
      SEE:    Paronychia fastigiata (Raf.) Fern. var. fastigiata

0915  1 SILENE STELLATA (L.) AIT. F. VAR. SCABRELLA NIEUWLAND.
      0915    PROBABLY NOT DISTINCT

0917  2 SILENE CUCUBALUS WIBEL.    SEE:    Silene vulgaris (Moench)
      Garcke

0924  2 LYCHNIS ALBA MILL.    SEE:    Silene alba (P. Mill.) Krause

0927  2 SAPONARIA VACCARIA L.    SEE:    Vaccaria pyramidata Medic.

0928  1 NELUMBO PENTAPETALA (WALT.) FERN.    SEE:    Nelumbo lutea
      (Willd.) Pers.

0930  1 NYMPHAEA TUBEROSA PAINE.    SEE:    Nymphaea odorata Ait.
      var. odorata

0931  1 NUPHAR ADVENA AIT.    SEE:    Nuphar luteum (L.) Sibthorp &
      Sm. ssp. variegatum (Dur.) E.O. Beal

0932  1 NUPHAR VARIEGATA ENGELM.        0932    PROBABLY NOT
      DISTINCT

0937  1 COPTIS GROENLANDICA (OEDER) FERN.    SEE:    Coptis trifolia
      (L.) Salisb. ssp. groenlandica (Oeder) Hulten

0938  1 ACTAEA ALBA (L.) MILL.        0938    PROBABLY NOT
      DISTINCT

0942  2 DELPHINIUM AJACIS L.    SEE:    Consolida ambigua (L.) Ball
      & Heywood

0949  1 ANEMONELLA THALICTROIDES (L.) SPACH.    SEE:    Thalictrum
      thalictroides (L.) Eames & Boivin

0950  1 HEPATICA ACUTILOBA DC.    SEE:    Hepatica nobilis P. Mill.
var. acuta (Pursh) Steyermark

0951  1 HEPATICA AMERICANA (DC.) KER.    SEE:    Hepatica nobilis P.
Mill. var. obtusa (Pursh) Steyermark

0958  1 RANUNCULUS TRICHOPHYLLUS CHAIX VAR. TYPICUS DREW.    SEE:
Ranunculus aquatilis L. var. capillaceus (Thuill.) DC.

0961  1 RANUNCULUS OBLONGIFOLIUS ELL.        0961    PROBABLY NOT
DISTINCT

0965  1 RANUNCULUS MICRANTHUS NUTT.    SEE:    Ranunculus micranthus
(Gray) Nutt. ex Torr. & Gray

0969  1 RANUNCULUS FASCICULARIS MUHL.    SEE:    Ranunculus
fascicularis Muhl. ex Bigelow

0971  2 RANUNCULUS REPENS L. VAR. VILLOSUS LAMOTTE.    SEE:
Ranunculus repens L. var. repens

0974  1 RANUNCULUS SEPTENTRIONALIS POIR. VAR. CARICETORUM (GREENE)
FERN.        0974    PROBABLY NOT DISTINCT

0978  1 THALICTRUM PERELEGANS GREENE.    SEE:    Thalictrum
pubescens Pursh var. pubescens

0984  1 CALYCOCARPUM LYONI (PURSH) NUTT.    SEE:    Calycocarpum
lyonii (Pursh) Gray

0988A 1 SASSAFRAS ALBIDUM (NUTT.) NEES. VAR. MOLLE (RAF.) FERN.
0988A   PROBABLY NOT DISTINCT

0989  1 BENZOIN AESTIVALE (L.) NEES.    SEE:    Lindera benzoin (L.)
Blume

0998  2 LEPIDIUM DRABA L.    SEE:    Cardaria draba (L.) Desv.

1000  2 LEPIDIUM DENSIFLORUM SCHRAD. VAR. TYPICUM THELLUNG.    SEE:
Lepidium densiflorum Schrad. var. densiflorum

1006  2 SISYMBRIUM THALIANUM (L.) J. GAY.    SEE:    Arabidopsis
thaliana (L.) Heynh.

1007  1 CAKILE EDENTULA (BIGEL.) HOOK. VAR. LACUSTRIS FERN.    SEE:
Cakile edentula (Bigelow) Hook. ssp. lacustris (Fern.) Hulten

1008  2 BRASSICA CAMPESTRIS L.   SEE:   Brassica rapa L. ssp. olifera DC.

1011  2 BRASSICA KABER (DC.) WHEELER VAR. PINNATIFIDA (STOKES) WHEELER.   SEE:   Sinapis arvensis L.

1016  1 RORIPPA PALUSTRIS (L.) BESS. VAR. GLABRATA (LUNELL) VICT. SEE:   Rorippa palustris (L.) Bess. ssp. glabra (O.E. Schulz) R. Stuckey var. glabrata (Lunell) R. Stuckey

1017  1 RORIPPA PALUSTRIS (L.) BESS. VAR. HISPIDA (DESV.) RYDB. SEE:   Rorippa palustris (L.) Bess. ssp. hispida (Desv.) Jonsell

1020  2 ARMORACIA RUSTICANA GAERTN.   SEE:   Armoracia rusticana (Lam.) Gaertn., Mey., & Scherb.

1024  1 CARDAMINE PRATENSIS L. VAR. PALUSTRIS WIMM. & GRAB.   SEE: Cardamine pratensis L.

1026  1 CARDAMINE PARVIFLORA L. VAR. ARENICOLA (BRITT.) O.E. SCHULZ.   SEE:   Cardamine parviflora L. ssp. parviflora var. arenicola (Britt.) O.E. Schulz

1027  1 DENTARIA LACINIATA MUHL.   SEE:   Dentaria laciniata Muhl. ex Willd.

1033  2 CAMELINA MICROCARPA ANDRZ.   SEE:   Camelina microcarpa Andrz. ex DC.

1034  1 DRABA BRACHYCARPA NUTT.   SEE:   Draba brachycarpa Nutt. ex Torr. & Gray

1035  2 DRABA VERNA L.   SEE:   Erophila verna (L.) Chev. ssp. verna

1037  2 DESCURAINIA BRACHYCARPA (RICHARDSON) O.E. SCHULZ.   SEE: Descurainia pinnata (Walt.) Britt. ssp. brachycarpa (Richards.) Detling

1038  1 ARABIS VIRGINICA (L.) POIR.   SEE:   Sibara virginica (L.) Rollins

1039  1 ARABIS PYCNOCARPA HOPKINS.   SEE:   Arabis hirsuta (L.) Scop. var. pycnocarpa (M. Hopkins) Rollins

1040  1 ARABIS PYCNOCARPA HOPKINS. VAR. ADPRESSIPILIS HOPKINS. SEE:   Arabis hirsuta (L.) Scop. var. adpressipilis (M. Hopkins) Rollins

1041  1 ARABIS VIRIDIS HARGER VAR. DEAMII HOPKINS.   SEE:   Arabis missouriensis Greene var. deamii (M. Hopkins) M. Hopkins

1043  1 ARABIS DENTATA T. & G.    SEE:    Arabis shortii (Fern.)
Gleason var. shortii

1049  1 ERYSIMUM ASPERUM DC.    SEE:    Erysimum asperum (Nutt.) DC.

1052  2 ALYSSUM ALYSSOIDES L.    SEE:    Alyssum alyssoides (L.) L.

1056  1 POLANISIA GRAVEOLENS RAF.    SEE:    Polanisia dodecandra
(L.) DC. ssp. dodecandra

1057  1 POLANISIA TRACHYSPERMA T. & G.    SEE:    Polanisia
dodecandra (L.) DC. ssp. trachysperma (Torr. & Gray) Iltis

1061  2 SEDUM ACRE L.        1061    PROBABLY NOT DISTINCT

1065  1 SULLIVANTIA OHIONIS T. & G.    SEE:    Sullivantia
sullivantii (Torr. & Gray) Britt.

1068  1 HEUCHERA AMERICANA L. VAR. BREVIPETALA ROSENDAHL, BUTTERS,
& LAKELA.    SEE:    Heuchera americana L. var. americana

1069  1 HEUCHERA AMERICANA L. VAR. INTERIOR ROSENDAHL, BUTTERS, &
LAKELA.    SEE:    Heuchera americana L. var. hirsuticaulis
(Wheelock) Rhosendahl, Butters & Lakela

1070  1 HEUCHERA AMERICANA L. VAR. HIRSUTICAULIS (WHEELOCK)
ROSENDAHL, BUTTERS, & LAKELA.        1070    PROBABLY NOT
DISTINCT

1071  1 HEUCHERA RICHARDSONII R. BR. VAR. AFFINIS ROSENDAHL,
BUTTERS, & LAKELA.    SEE:    Heuchera richardsonii R. Br.

1072  1 HEUCHERA RICHARDSONII R. BR. VAR. GRAYANA ROSENDAHL,
BUTTERS, & LAKELA.        1072    PROBABLY NOT DISTINCT

1073  1 HEUCHERA VILLOSA MICHX. VAR. MACRORHIZA (SMALL) ROSENDAHL,
BUTTERS, & LAKELA.    SEE:    Heuchera villosa Michx.

1074  1 HEUCHERA PARVIFLORA BARTL. VAR. RUGELII (SHUTTLW. APUD
KUNTZE) ROSENDAHL, BUTTERS, & LAKELA.    SEE:    Heuchera
parviflora Nutt. ex Torr. & Gra

1076  1 CHRYSOSPLENIUM AMERICANUM SCHWEIN.    SEE:    Chrysosplenium
americanum Schwein. ex Hook.

1079  1 RIBES AMERICANUM MILL.        1079    PROBABLY NOT
DISTINCT

1080  1 GROSSULARIA CYNOSBATI (L.) MILL.   SEE:   Ribes cynosbati L.

1081  1 GROSSULARIA MISSOURIENSIS (NUTT.) COV. & BRITT.   SEE: Ribes missouriense Nutt. ex Torr. & Gray

1082  1 GROSSULARIA HIRTELLA (MICHX.) SPACH.   SEE:   Ribes hirtellum Michx.

1091  1 GILLENIA STIPULATA (MUHL.) TREL.   SEE:   Porteranthus stipulatus (Muhl. ex Willd.) Britt.

1092  1 MALUS CORONARIA (L.) MILL.   SEE:   Malus coronaria (L.) P. Mill. var. coronaria

1100  1 AMELANCHIER LAEVIS WIEG.   SEE:   Amelanchier arborea (Michx. f.) Fern. var. laevis (Wieg.) Ahles

1102  1 CRATAEGUS PYRACANTHOIDES BEADLE VAR. ARBOREA (BEADLE) PALMER.         1102    PROBABLY NOT DISTINCT

1103  1 CRATAEGUS REGALIS BEADLE.         1103    PROBABLY NOT DISTINCT

1104  1 CRATAEGUS ACUTIFOLIA SARG.         1104    PROBABLY NOT DISTINCT

1106  1 CRATAEGUS COLLINA CHAPM.         1106    PROBABLY NOT DISTINCT

1112  1 CRATAEGUS RUBELLA BEADLE.         1112    PROBABLY NOT DISTINCT

1116  1 CRATAEGUS GATTINGERI ASHE.         1116    PROBABLY NOT DISTINCT

1117  1 CRATAEGUS PLATYCARPA SARG.         1117    PROBABLY NOT DISTINCT

1118  1 CRATAEGUS RUGOSA ASHE.         1118    PROBABLY NOT DISTINCT

1121  1 CRATAEGUS PUTNAMIANA SARG.   SEE:   Crataegus chrysocarpa Ashe

1126  1 CRATAEGUS SUCCULENTA SCHRADER.   SEE:   Crataegus succulenta Schrad. ex Link

1127  1 CRATAEGUS INCAEDUA SARG.   SEE:   Crataegus X incaedua
Sarg.

1131  1 RUBUS IDAEUS L. VAR. CANADENSIS RICHARDSON.   SEE:   Rubus
idaeus L. ssp. sachalinensis (Levl.) Focke

1138  1 RUBUS ALLEGHENIENSIS PORTER.   SEE:   Rubus allegheniensis
Porter ex Bailey

1139  1 RUBUS IMPOS BAILEY.   SEE:   Rubus alumnus Bailey

1147  1 FRAGARIA VIRGINIANA DUCHESNE VAR. ILLINOENSIS (PRINCE)
GRAY.   SEE:   Fragaria virginiana Duchesne ssp. platypetala
(Rydb.) Staudt

1155  1 POTENTILLA MONSPELIENSIS L.   SEE:   Potentilla norvegica
L. ssp. monspeliensis (L.) Aschers & Graebn.

1157  1 POTENTILLA SIMPLEX MICHX. VAR. TYPICA FERN.   SEE:
Potentilla simplex Michx. var. simplex

1164  1 GEUM ALEPPICUM JACQ. VAR. STRICTUM (AIT.) FERN.   SEE:
Geum aleppicum Jacq.

1175  2 ROSA RUBIGINOSA L.   SEE:   Rosa eglanteria L.

1180  1 ROSA SUFFULTA GREENE.   SEE:   Rosa arkansana Porter var.
suffulta (Greene) Cockerell

1183  1 PRUNUS LANATA (SUDW.) MACK. & BUSH.   SEE:   Prunus nigra
Ait.

1186  1 PRUNUS NIGRA AIT.        1186   PROBABLY NOT DISTINCT

1192  2 DESMANTHUS ILLINOENSIS (MICHX.) MACM.   SEE:   Desmanthus
illinoensis (Michx.) MacM. ex B.L. Robins. & Fern.

1195  1 CASSIA NICTITANS L. VAR. LEIOCARPA FERN.        1195
PROBABLY NOT DISTINCT

1197  1 CASSIA FASCICULATA MICHX. VAR. ROBUSTA (POLLARD) MACBRIDE.
1197   PROBABLY NOT DISTINCT

1203  1 GLEDITSIA TEXANA SARG.   SEE:   Gleditsia X texana Sarg.

1205  1 CLADRASTIS LUTEA (MICHX. F.) KOCH.    SEE:    Cladrastis
      kentukea (Dum.-Cours.) Rudd

1206  1 BAPTISIA LEUCOPHAEA NUTT.          1206    PROBABLY NOT
      DISTINCT

1207  1 BAPTISIA AUSTRALIS (L.) R. BR.        1207    PROBABLY
      NOT DISTINCT

1209  1 BAPTISIA LEUCANTHA T. & G.    SEE:    Baptisia lactea (Raf.)
      Thieret

1214  2 MELILOTUS ALBA DESR.    SEE:    Melilotus alba Medic.

1215  2 MELILOTUS OFFICINALIS (L.) LAM.    SEE:    Melilotus
      officinalis (L.) Pallas

1216  2 TRIFOLIUM ARVENSE L.          1216    PROBABLY NOT DISTINCT

1217  2 TRIFOLIUM PRATENSE L.          1217    PROBABLY NOT
      DISTINCT

1218  2 TRIFOLIUM REPENS L.          1218    PROBABLY NOT DISTINCT

1219  1 TRIFOLIUM REFLEXUM L. VAR. GLABRUM LOJACONO.    SEE:
      Trifolium reflexum L.

1221  2 TRIFOLIUM AGRARIUM L.    SEE:    Trifolium aureum Pollich.

1225  1 AMORPHA CANESCENS NUTT.    SEE:    Amorpha canescens Pursh

1227  1 PETALOSTEMUM PURPUREUM (VENT.) RYDB.    SEE:    Dalea
      purpurea Vent.

1228  1 PETALOSTEMUM CANDIDUM (WILLD.) MICHX.    SEE:    Dalea
      candida (Michx.) Willd.

1230  1 TEPHROSIA VIRGINIANA (L.) PERS. VAR. HOLOSERICEA (NUTT.)
      T. & G.          1230    PROBABLY NOT DISTINCT

1234  2 CORONILLA VARIA L.          1234    PROBABLY NOT DISTINCT

1236  1 DESMODIUM ROTUNDIFOLIUM (MICHX.) DC.    SEE:    Desmodium rotundifolium DC.

1240  1 DESMODIUM ACUMINATUM (MICHX.) DC.    SEE:    Desmodium glutinosum (Muhl. ex Willd.) Wood

1243  1 DESMODIUM BRACTEOSUM (MICHX.) DC.    SEE:    Desmodium cuspidatum (Muhl. ex Willd.) Loud. var. cuspidatum

1244  1 DESMODIUM BRACTEOSUM (MICHX.) DC. VAR. LONGIFOLIUM (T. & G.) ROB.    SEE:    Desmodium cuspidatum (Muhl. ex Willd.) Loud. var. longifolium (Torr. & Gray) Schub.

1248  1 DESMODIUM DILLENII DARL.    SEE:    Desmodium perplexum Schub.

1251  1 DESMODIUM CILIARE DC.    SEE:    Desmodium ciliare (Muhl. ex Willd.) DC.

1252  1 DESMODIUM RIGIDUM (ELL.) DC.    SEE:    Desmodium obtusum (Muhl. ex Willd.) DC.

1253  2 LESPEDEZA STRIATA (THUNB.) H. & A.    SEE:    Kummerowia striata (Thunb.) Schindl.

1254  2 LESPEDEZA STIPULACEA MAXIM.    SEE:    Kummerowia stipulacia (Maxim.) Makino

1256  1 LESPEDEZA NUTTALLII DARL.    SEE:    Lespedeza X nuttallii Darl.

1259  1 LESPEDEZA VIRGINICA (L.) DESV. FORMA DEAMII HOPKINS. SEE:    Lespedeza virginica (L.) Britt. f. deamii Hopkins

1268  1 VICIA AMERICANA MUHL.    SEE:    Vicia americana Muhl. ex Willd.

1272  1 LATHYRUS VENOSUS MUHL. VAR. INTONSUS BUTTERS & ST. JOHN. SEE:    Lathyrus venosus Muhl. ex Willd. ssp. venosus var. intonsus (Butters & St. John)

1273  1 LATHYRUS PALUSTRIS L.           1273    PROBABLY NOT DISTINCT

1274  1 LATHYRUS PALUSTRIS L. VAR. LINEARIFOLIUS SER.    SEE: Lathyrus palustris L. var. palustris

1278  1 AMPHICARPA BRACTEATA L. FERN. VAR. COMOSA (L.) FERN. 1278    PROBABLY NOT DISTINCT

1279  1 APIOS AMERICANA MEDIC.      1279    PROBABLY NOT
DISTINCT

1280  1 GALACTIA VOLUBILIS (L.) BRITT. VAR. MISSISSIPPIENSIS VAIL.
1280   PROBABLY NOT DISTINCT

1282  1 STROPHOSTYLES HELVOLA (L.) BRITT.    SEE:   Strophostyles
helvola (L.) Ell.

1283  1 STROPHOSTYLES UMBELLATA (MUHL.) BRITT.    SEE:
Strophostyles umbellata (Muhl. ex Willd.) Britt.

1290  1 GERANIUM CAROLINIANUM VAR. CONFERTIFLORUM FERN.    SEE:
Geranium carolinianum L. var. confertiflorum Fern.

1291  2 GERANIUM PUSILLUM BURM. F.    SEE:   Geranium pusillum L.

1296  1 OXALIS FLORIDA SALISB.    SEE:   Oxalis dillenii Jacq. ssp.
filipes (Small) Eiten

1298  1 OXALIS EUROPAEA JORDAN FORMA CYMOSA (SMALL) WIEG.    SEE:
Oxalis europaea Jord. f. cymosa (Small) Wieg.

1304  1 LINUM MEDIUM (PLANCH.) TREL. VAR. TEXANUM (PLANCH.) FERN.
SEE:   Linum medium (Planch.) Britt. var. texanum (Planch.)
Fern.

1306  1 PTELEA TRIFOLIATA L.    SEE:   Ptelea trifoliata L. ssp.
trifoliata

1314  1 POLYGALA AMBIGUA NUTT.    SEE:   Polygala verticillata L.
var. ambigua (Nutt.) Wood

1324  1 ACALYPHA RHOMBOIDEA RAF. VAR. DEAMII WEATHERBY.    SEE:
Acalypha deamii (Weatherby) Ahles

1329  1 EUPHORBIA POLYGONIFOLIA L.    SEE:   Chamaesyce
polygonifolia (L.) Small

1330  1 EUPHORBIA SERPENS HBK.    SEE:   Chamaesyce serpens
(H.B.K.) Small

1331  1 EUPHORBIA GLYPTOSPERMA ENGELM.    SEE:   Chamaesyce
glyptosperma (Engelm.) Small

1332  1 EUPHORBIA HUMISTRATA ENGELM.    SEE:   Chamaesyce
humistrata (Engelm. ex Gray) Small

1334 1 EUPHORBIA VERMICULATA RAF.    SEE:    Chamaesyce vermiculata (Raf.) House

1335 1 EUPHORBIA SUPINA RAF.    SEE:    Chamaesyce maculata (L.) Small

1337 1 EUPHORBIA DENTATA MICHX.    SEE:    Poinsettia dentata (Michx.) Klotzsch & Garcke

1338 1 EUPHORBIA HETEROPHYLLA L.    SEE:    Poinsettia heterophylla (L.) Klotzsch & Garcke

1344 1 CALLITRICHE AUSTINI ENGELM.    SEE:    Callitriche terrestris Raf. emend. Torr.

1345 1 CALLITRICHE HETEROPHYLLA PURSH.    SEE:    Callitriche heterophylla Pursh emend. Darby

1348 1 RHUS VERNIX L.    SEE:    Toxicodendron vernix (L.) Kuntze

1352 1 RHUS RADICANS L. VAR. LITTORALIS (MEARNS) DEAM, COMB. NOV. SEE:    Toxicodendron radicans (L.) Kuntze ssp. radicans

1363 1 ACER NEGUNDO L. VAR. VIOLACEUM KIRCHNER.    SEE:    Acer negundo L. var. violaceum J. Miller

1368 1 ACER SACCHARUM MARSH. FORMA SCHNECKII (REHDER) DEAM, COMB. NOV.    SEE:    Acer saccharum Marsh. f. walpolei (Rehd.) Deam

1370 1 AESCULUS OCTANDRA MARSH.    SEE:    Aesculus flava Soland.

1371 1 IMPATIENS BIFLORA WALT.    SEE:    Impatiens capensis Meerb.

1377 1 CEANOTHUS OVATUS DESF.    SEE:    Ceanothus herbaceus Raf.

1380 1 VITIS CINEREA ENGELM.    SEE:    Vitis cinerea Engelm. ex Millard

1385 1 PARTHENOCISSUS VITACEA (KNERR) HITCHC.    SEE:    Parthenocissus inserta (Kern.) Fritsch

1390   2 MALVA SYLVESTRIS L. VAR. MAURETIANA (L.) BOISS.   SEE:
Malva sylvestris L.

1397   1 HIBISCUS MILITARIS CAV.   SEE:   Hibiscus laevis All.

1399   1 HIBISCUS PALUSTRIS L.         1399   PROBABLY NOT
DISTINCT

1402   1 ASCYRUM HYPERICOIDES L. VAR. MULTICAULE (MICHX.) FERN.
SEE:   Hypericum stragulum P. Adams & Robson

1403   1 HYPERICUM ASCYRON L.   SEE:   Hypericum pyramidatum Ait.

1419   1 HYPERICUM VIRGINICUM L.   SEE:   Triandenum virginicum
(L.) Raf.

1420   1 HYPERICUM VIRGINICUM L. VAR. FRASERI (SPACH) FERN.   SEE:
Triandenum fraseri (Spach) Gleason

1421   1 HYPERICUM TUBULOSUM WALT.   SEE:   Triandenum tubulosum
(Walt.) Gleason

1422   1 HYPERICUM TUBULOSUM WALT. VAR. WALTERI (GMEL.) LOTT.
SEE:   Triandenum walteri (J.G. Gmel.) Gleason

1428   1 LECHEA RACEMULOSA LAM.   SEE:   Lechea racemulosa Michx.

1431   1 LECHEA LEGGETTII BRITTON & HOLLICK VAR. MONILIFORMIS
(BICKNELL) HODGDON.   SEE:   Lechea pulchella Raf. var.
moniliformis (Bickn.) Seymour

1432   1 HYBANTHUS CONCOLOR (FORST.) SPRENG.   SEE:   Hybanthus
concolor (T.F. Forst.) Spreng.

1436   1 VIOLA CUCULLATA AIT.   SEE:   Viola obliqua Hill

1438   1 VIOLA PAPILIONACEA PURSH.   SEE:   Viola sororia Willd.

1440   1 VIOLA SORORIA WILLD.         1440   PROBABLY NOT DISTINCT

1445   1 VIOLA PALLENS (BANKS) BRAINERD.         1445   PROBABLY
NOT DISTINCT

1446  1 VIOLA BLANDA WILLD.    SEE:    Viola macloskeyi Lloyd ssp.
pallens (Banks ex DC.) M.S. Baker

1447  1 VIOLA INCOGNITA BRAINERD VAR. FORBESII BRAINERD.    SEE:
Viola incognita Brainerd

1448  1 VIOLA ERIOCARPA SCHWEIN.    SEE:    Viola pubescens Ait.
var. eriocarpa (Schwein.) Russell

1451  2 VIOLA KITAIBELIANA ROEM. & SCHULTES VAR. RAFINESQUII
(GREENE) FERN.    SEE:    Viola rafinesquii Greene

1458  1 OPUNTIA HUMIFUSA RAF.    SEE:    Opuntia humifusa (Raf.)
Raf.

1461  1 ROTALA RAMOSIOR (L.) KOEHNE VAR. TYPICA FERN. & GRISC.
SEE:    Rotala ramosior (L.) Koehne

1462  1 ROTALA RAMOSIOR (L.) KOEHNE VAR. INTERIOR FERN. & GRISC.
1462   PROBABLY NOT DISTINCT

1467  1 CUPHEA PETIOLATA (L.) KOEHNE.    SEE:    Cuphea viscosissima
Jacq.

1471  1 RHEXIA MARIANA L. VAR. LEIOSPERMA FERN. & GRISC.    SEE:
Rhexia mariana L. var. mariana

1472  1 JUSSIAEA DECURRENS (WALT.) DC.    SEE:    Ludwigia decurrens
Walt.

1473  1 JUSSIAEA DIFFUSA FORSKAL.    SEE:    Ludwigia peploides
(H.B.K.) Raven ssp. glabrescens (Kuntze) Raven

1474  1 LUDWIGIA PALUSTRIS (L.) ELL. VAR. AMERICANA (DC.) FERN. &
GRESC.    SEE:    Ludwigia palustris (L.) Ell.

1478  1 LUDWIGIA SPHAEROCARPA ELL. VAR. DEAMII FERN. & GRISC.
SEE:    Ludwigia sphaerocarpa Ell.

1479  1 EPILOBIUM ANGUSTIFOLIUM L.    SEE:    Epilobium
angustifolium L. ssp. angustifolium

1481  1 EPILOBIUM DENSUM RAF.        1481    PROBABLY NOT
DISTINCT

1482  1 EPILOBIUM COLORATUM MUHL.    SEE:    Epilobium coloratum
Biehler

1483  1 EPILOBIUM GLANDULOSUM LEHM. VAR. ADENOCAULON (HAUSSK.)
      FERN.   SEE:   Epilobium ciliatum Raf. ssp. ciliatum

1484  1 OENOTHERA PYCNOCARPA ATKINSON & BARTLETT.   SEE:
      Oenothera biennis L. ssp. biennis

1485  1 OENOTHERA NUTANS ATKINSON & BARTLETT.          1485
      PROBABLY NOT DISTINCT

1486  1 OENOTHERA CANOVIRENS STEELE.   SEE:   Oenothera villosa
      Thunb. ssp. canovirens (Steele) D. Dietr. & Raven

1490  1 OENOTHERA TETRAGONA ROTH. VAR. LONGISTIPATA (PENNELL)
      MUNZ.   SEE:   Oenothera fruticosa L. ssp. fruticosa

1491  1 OENOTHERA PERENNIS L. VAR. TYPICA MUNZ.   SEE:   Oenothera
      perennis L.

1498  1 CIRCAEA LATIFOLIA HILL.   SEE:   Circaea lutetiana (L.)
      Aschers. & Magnus ssp. canadensis (L.) Aschers. & Magnus

1503  1 MYRIOPHYLLUM VERTICILLATUM L. VAR. PECTINATUM WALLR.
      SEE:   Myriophyllum verticillatum L.

1521  1 CHAEROPHYLLUM PROCUMBENS (L.) CRANTZ.   SEE:
      Chaerophyllum procumbens (L.) Crantz var. procumbens

1526  1 OSMORHIZA LONGISTYLIS (TORR.) DC. VAR. VILLICAULIS FERN.
      1526   PROBABLY NOT DISTINCT

1527  1 OSMORHIZA LONGISTYLIS (TORR.) DC. VAR. BRACHYCOMA BLAKE.
      1527   PROBABLY NOT DISTINCT

1539  1 PERIDERIDIA AMERICANA (NUTT.) REICHENB.   SEE:
      Perideridia americana (Nutt. ex DC.) Reichenb.

1542  1 THASPIUM TRIFOLIATUM (L.) BRITT. VAR. FLAVUM BLAKE.   SEE:
      Thaspium trifoliatum (L.) Gray var. flavum Blake

1545  1 ANGELICA VILLOSA (WALT.) BSP.   SEE:   Angelica venenosa
      (Greenway) Fern.

1552  1 NYSSA SYLVATICA MARSH. VAR. TYPICA FERN.   SEE:   Nyssa
      sylvatica Marsh. var. sylvatica

1553  1 NYSSA SYLVATICA MARSH. VAR. CAROLINIANA (POIR.) FERN.
      1553   PROBABLY NOT DISTINCT

1558  1 CORNUS STOLONIFERA MICHX. VAR. BAILEYI (COULTER & EVANS)
DRESCHER.    SEE:    Cornus sericea L. ssp. sericea

1560  1 CORNUS STOLONIFERA MICHX.         1560    PROBABLY NOT
DISTINCT

1561  1 CORNUS RACEMOSA LAM.    SEE:    Cornus foemina P. Mill. ssp.
racemosa (Lam.) J.S. Wilson

1562  1 CORNUS STRICTA LAM.    SEE:    Cornus foemina P. Mill. ssp.
foemina

1563  1 CORNUS OBLIQUA RAF.    SEE:    Cornus amomum P. Mill. ssp.
obliqua (Raf.) J.S. Wilson

1564  1 CORNUS AMOMUM MILL.    SEE:    Cornus amomum P. Mill. ssp.
amomum

1566  1 CHIMAPHILA UMBELLATA (L.) BART. VAR. CISATLANTICA BLAKE.
SEE:    Chimaphila umbellata (L.) Bart. ssp. cisatlantica
(Blake) Hulten

1567  1 PYROLA SECUNDA L.    SEE:    Orthilia secunda (L.) House
ssp. secunda

1570  1 PYROLA ROTUNDIFOLIA L. VAR. AMERICANA (SWEET) FERN.    SEE:
Pyrola americana Sweet

1571  1 PYROLA ASARIFOLIA MICHX. VAR. INCARNATA (FISCH.) FERN.
SEE:    Pyrola asarifolia Michx. var. purpurea (Bunge) Fern.

1573  1 MONOTROPA HYPOPITYS L. VAR. RUBRA (TORR.) FARW.    SEE:
Monotropa hypopithys L.

1575  1 ANDROMEDA GLAUCOPHYLLA LINK.    SEE:    Andromeda polifolia
L. var. glaucophylla (Link.) DC.

1580  1 ARCTOSTAPHYLOS UVA-URSI (L.) SPRENG. VAR. COACTILIS FERN.
& MACB.    SEE:    Arctostaphylos uva-ursi (L.) Spreng. ssp.
coactilis (Fern. & Macbr.) Love, Love & Kapoor

1583  1 VACCINIUM STAMINEUM L. VAR. NEGLECTUM (SMALL) DEAM.
1583    PROBABLY NOT DISTINCT

1587  1 VACCINIUM VACILLANS KALM EX TORREY.    SEE:    Vaccinium
pallidum Ait.

1588  1 VACCINIUM CANADENSE KALM.    SEE:    Vaccinium myrtilloides
Michx.

1593  1 SAMOLUS PAUCIFLORUS RAF.   SEE:   Samolus valerandi L.
ssp. parviflorus (Raf.) Hulten

1601  1 LYSIMACHIA LONGIFOLIA PURSH.   SEE:   Lysimachia
quadriflora Sims

1604  1 CENTUNCULUS MINIMUS L.   SEE:   Anagallis minima (L.)
Krause

1610  1 FRAXINUS BILTMOREANA BEADLE.   SEE:   Fraxinus americana
L. var. biltmoreana (Beadle) J. Wright ex Fern.

1611  1 FRAXINUS LANCEOLATA BORKH.   SEE:   Fraxinus pennsylvanica
Marsh.

1612  1 FRAXINUS PENNSYLVANICA MARSH.        1612   PROBABLY NOT
DISTINCT

1613  1 FRAXINUS TOMENTOSA MICHX. F.   SEE:   Fraxinus profunda
(Bush) Bush

1615  1 FRAXINUS NIGRA L.   SEE:   Fraxinus nigra Marsh.

1619  1 SABATIA CAMPANULATA (L.) TORR. VAR. GRACILIS (MICHX.)
FERN.   SEE:   Sabatia campanulata (L.) Torr.

1622  1 GENTIANA CRINITA FROEL.   SEE:   Gentianopsis crinita
(Froel.) Ma

1623  1 GENTIANA PROCERA HOLM.   SEE:   Gentianopsis procera
(Holm) Ma

1624  1 GENTIANA QUINQUEFOLIA L. VAR. OCCIDENTALIS (GRAY) HITCHC.
SEE:   Gentianella quinquefolia (L.) Small ssp. occidentalis
(Gray) J. Gillett

1627  1 GENTIANA PUBERULA MICHX.   SEE:   Gentiana puberulenta J.
Pringle

1628  1 GENTIANA FLAVIDA GRAY.   SEE:   Gentiana alba Muhl.

1634  1 APOCYNUM ANDROSAEMIFOLIUM L.   SEE:   Apocynum
androsaemifolium L. ssp. androsaemifolium

1635  1 APOCYNUM MEDIUM GREENE.   SEE:   Apocynum X medium Greene

1636 1 APOCYNUM MEDIUM GREENE VAR. SARNIENSE (GREENE) WOODSON.
1636    PROBABLY NOT DISTINCT

1637 1 APOCYNUM MEDIUM GREENE VAR. LEUCONEURON (GREENE) WOODSON.
1637    PROBABLY NOT DISTINCT

1638 1 APOCYNUM CANNABINUM L.    SEE:    Apocynum cannabinum L.
var. cannabinum

1639 1 APOCYNUM CANNABINUM L. VAR. PUBESCENS (MITCHELL) A. DC.
1639    PROBABLY NOT DISTINCT

1641 1 APOCYNUM SIBIRICUM JACQ.    SEE:   Apocynum cannabinum L.
var. hypericifolium Gray

1642 1 APOCYNUM SIBIRICUM JACQ. VAR. FARWELLII (GREENE) FERN.
1642    PROBABLY NOT DISTINCT

1643 1 ACERATES HIRTELLA PENNELL.    SEE:    Asclepias hirtella
(Pennell) Woods.

1644 1 ACERATES VIRIDIFLORA (RAF.) EATON.    SEE:    Asclepias
viridiflora Raf.

1647 1 ASCLEPIAS MEADII TORR.    SEE:    Asclepias meadii Torr. ex
Gray

1649 1 ASCLEPIAS SULLIVANTII ENGELM.    SEE:    Asclepias
sullivantii Engelm. ex Gray

1650 1 ASCLEPIAS INCARNATA L.    SEE:    Asclepias incarnata L.
ssp. incarnata

1654 1 ASCLEPIAS PHYTOLACCOIDES PURSH.    SEE:    Asclepias
exaltata L.

1657 1 AMPELAMUS ALBIDUS (NUTT.) BRITT.    SEE:    Cynanchum laeve
(Michx.) Pers.

1658 1 GONOLOBUS GONOCARPOS (WALT.) PERRY.    SEE:    Matelea
gonocarpa (Walt.) Shinners

1659 1 GONOLOBUS OBLIQUUS (JACQ.) SCHULTES.    SEE:    Matelea
obliqua (Jacq.) Woods.

1669 1 CONVOLVULUS SPITHAMAEUS L.    SEE:    Calystegia spithamaea
(L.) Pursh ssp. spithamaea

1671   1 CONVOLVULUS SEPIUM L.   SEE:   Calystegia sepium (L.) R.
Br. ssp. sepium

1672   1 CONVOLVULUS SEPIUM L. VAR. FRATERNIFLORUS MACK. & BUSH.
SEE:   Calystegia fraterniflora (Mackenzie & Bush) Brummitt

1673   1 CONVOLVULUS REPENS L.   SEE:   Calystegia sepium (L.) R.
Br. ssp. americana (Sims) Brummitt

1677   1 IPOMOEA PANDURATA (L.) G.F.W. MEY. VAR. RUBESCENS CHOISY.
1677   PROBABLY NOT DISTINCT

1678   2 IPOMOEA HEDERACEA JACQ.   SEE:   Ipomoea nil (L.) Roth

1679   2 IPOMOEA PURPUREA (L.) ROTH.        1679   PROBABLY NOT
DISTINCT

1680   2 QUAMOCLIT COCCINEA (L.) MOENCH.   SEE:   Ipomoea coccinea
L.

1684   1 PHLOX CAROLINA L. VAR. TRIFLORA (MICHX.) WHERRY.   SEE:
Phlox glaberrima L. ssp. triflora (Michx.) Wherry

1685   1 PHLOX GLABERRIMA L.   SEE:   Phlox glaberrima L. ssp.
glaberrima

1698   1 PHACELIA COVILLEI WATS.   SEE:   Phacelia ranunculacea
(Nutt.) Constance

1707   1 MYOSOTIS VIRGINICA (L.) BSP.   SEE:   Myosotis verna Nutt.

1708   1 MYOSOTIS VIRGINICA (L.) BSP. VAR. MACROSPERMA (ENGELM.)
FERN.   SEE:   Myosotis macrosperma Engelm.

1709   2 MYOSOTIS MICRANTHA PALL.   SEE:   Myosotis stricta Link ex
Roemer & Schultes

1710   1 MERTENSIA VIRGINICA (L.) LINK.   SEE:   Mertensia
virginica (L.) Pers. ex Link

1711   2 LITHOSPERMUM ARVENSE L.   SEE:   Buglossoides arvense (L.)
I.M. Johnston

1715   1 LITHOSPERMUM CROCEUM FERN.   SEE:   Lithospermum
caroliniense (J.F. Gmel.) MacM.

1718 1 VERBENA CANADENSIS (L.) BRITT.    SEE:    Glandularia
canadensis (L.) Nutt.

1728 1 TEUCRIUM OCCIDENTALE GRAY.    SEE:    Teucrium canadense L.
var. boreale (Bickn.) Shinners

1729 1 TEUCRIUM OCCIDENTALE GRAY VAR. BOREALE (BICKN.) FERN.
1729    PROBABLY NOT DISTINCT

1730 1 ISANTHUS BRACHIATUS (L.) BSP.    SEE:    Trichostema
brachiatum L.

1738 1 SCUTELLARIA INCANA SPRENG.    SEE:    Scutellaria incana
Biehler

1739 1 SCUTELLARIA OVALIFOLIA PERS.    SEE:    Scutellaria
elliptica Muhl. var. elliptica

1746 2 GLECOMA HEDERACEA L. VAR. PARVIFLORA (BENTH.) HOUSE.
SEE:    Glechoma hederacea L. var. micrantha Moric.

1747 1 PRUNELLA VULGARIS L. VAR. LANCEOLATA (BART.) FERN.    SEE:
Prunella vulgaris L. var. elongata Benth.

1748 1 PHYSOSTEGIA SPECIOSA (SWEET) SWEET.    SEE:    Physostegia
virginiana (L.) Benth.

1749 1 PHYSOSTEGIA VIRGINIANA (L.) BENTH.          1749
PROBABLY NOT DISTINCT

1750 1 SYNANDRA HISPIDULA (MICHX.) BRITT.    SEE:    Synandra
hispidula (Michx.) Baill.

1756 1 STACHYS HISPIDA PURSH.    SEE:    Stachys tenuifolia Willd.
var. hispida (Pursh.) Fern.

1759 1 STACHYS RIDDELLII HOUSE.    SEE:    Stachys nuttallii
Shuttlw. ex Benth.

1763 2 SALVIA SYLVESTRIS L.    SEE:    Salvia nemorosa L.

1768 1 MONARDO PUNCTATA L. VAR. VILLICAULIS PENNELL.    SEE:
Monarda punctata L. var. villicaulis (Pennell) Shinners

1769 1 BLEPHILIA CILIATA (L.) RAF.    SEE:    Blephilia ciliata
(L.) Benth.

1772  1 HEDEOMA HISPIDA PURSH.    SEE:    Hedeoma hispidum Pursh

1774  1 SATUREJA VULGARIS (L.) FRITSCH.    SEE:    Clinopodium
      vulgare L.

1775  1 SATUREJA GLABRA (NUTT.) FERN.    SEE:    Calamintha
      arkansana (Nutt.) Shinners

1783  1 LYCOPUS SESSILIFOLIUS GRAY.    SEE:    Lycopus amplectens
      Raf.

1785  1 LYCOPUS AMERICANUS MUHL.    SEE:    Lycopus americanus Muhl.
      ex Bart.

1786  1 LYCOPUS AMERICANUS MUHL. VAR. LONGII BENNER.          1786
      PROBABLY NOT DISTINCT

1788  2 MENTHA PIPERITA L.    SEE:    Mentha X piperita L.

1789  2 MENTHA LONGIFOLIA (L.) HUDS. VAR. MOLLISSIMA BORKH.
      1789    PROBABLY NOT DISTINCT

1790  2 MENTHA ROTUNDIFOLIA L.    SEE:    Mentha X rotundifolia (L.)
      Huds.

1792  1 MENTHA ARVENSIS L. VAR. SATIVA BENTH.    SEE:    Mentha
      canadensis L.

1795  2 NICANDRA PHYSALODES (L.) PERS.    SEE:    Nicandra
      physalodes (L.) Gaertn.

1796  2 LYCIUM HALIMIFOLIUM MILL.    SEE:    Lycium barbarum L.

1799  1 PHYSALIS NYCTAGINEA DUNAL.    SEE:    Physalis heterophylla
      Nees var. heterophylla

1801  1 PHYSALIS PRUINOSA L.    SEE:    Physalis pubescens L. var.
      integrifolia (Dunal) Waterfall

1802  1 PHYSALIS HETEROPHYLLA NEES.          1802    PROBABLY NOT
      DISTINCT

1803  1 PHYSALIS AMBIGUA (A. GRAY) RYDB.          1803    PROBABLY
      NOT DISTINCT

1805  2 SOLANUM ROSTRATUM DUNAL.    SEE:    Solanum cornutum Lam.

1820  1 CHELONE GLABRA L. VAR. TYPICA PENNELL.    SEE:    Chelone
glabra L. var. glabra

1822  1 PENSTEMON TUBAEFLORUS NUTT.    SEE:    Penstemon tubiflorus
Nutt.

1827  1 PENSTEMON CANESCENS (BRITT.) BRITT. VAR. TYPICUS PENNELL.
SEE:    Penstemon canescens (Britt.) Britt.

1836  1 LINDERNIA DUBIA (L.) PENNELL VAR. TYPICA PENNELL.    SEE:
Lindernia dubia (L.) Pennell var. dubia

1837  1 LINDERNIA DUBIA (L.) PENNELL VAR. MAJOR (PURSH) PENNELL.
1837   PROBABLY NOT DISTINCT

1838  1 LINDERNIA ANAGALLIDEA (MICHX.) PENNELL.    SEE:    Lindernia
dubia (L.) Pennell var. anagallidea (Michx.) Cooperrider

1839  1 VERONICA PEREGRINA L. VAR. TYPICA PENNELL.    SEE:
Veronica peregrina L. ssp. peregrina

1840  1 VERONICA PEREGRINA L. VAR. XALAPENSIS (HBK.) PENNELL.
SEE:    Veronica peregrina L. ssp. xalapensis (H.B.K.) Pennell

1845  2 VERONICA CHAMAEDRYS L.    SEE:    Veronica catenata Pennell

1847  1 VERONICA AMERICANA (RAF.) SCHWEIN.    SEE:    Veronica
americana (Raf.) Schwein. ex Benth.

1848  1 VERONICA GLANDIFERA PENNELL.    SEE:    Veronica cymbalaria
Bodard

1849  1 VERONICA CONNATA RAF. VAR. TYPICA PENNELL.          1849
PROBABLY NOT DISTINCT

1853  1 GERARDIA PURPUREA L.    SEE:    Agalinis purpurea (L.)
Pennell

1854  1 GERARDIA PAUPERCULA (GRAY) BRITT. VAR. TYPICA PENNELL.
SEE:    Agalinis paupercula (Gray) Britt.

1855  1 GERARDIA PAUPERCULA (GRAY) BRITT. VAR. BOREALIS (PENNELL)
PENNELL.    SEE:    Agalinis paupercula (Gray) Britt. var.
borealis Pennell

1856  1 GERARDIA TENUIFOLIA VAHL VAR. TYPICA PENNELL.   SEE:
Agalinis tenuifolia (Vahl) Raf. var. tenuifolia

1857  1 GERARDIA TENUIFOLIA VAHL VAR. MACROPHYLLA BENTH.   SEE:
Agalinis besseyana Britt.

1858  1 GERARDIA TENUIFOLIA VAHL VAR. PARVIFLORA NUTT.   SEE:
Agalinis tenuifolia (Vahl) Raf. var. parviflora (Nutt.)
Pennell

1859  1 GERARDIA SKINNERIANA WOOD.   SEE:   Agalinis skinneriana
(Wood) Britt.

1860  1 GERARDIA GATTINGERI SMALL.   SEE:   Agalinis gattingeri
(Small) Small

1862  1 AUREOLARIA FLAVA (L.) FARW. VAR. TYPICA PENNELL.   SEE:
Aureolaria flava (L.) Farw. var. flava

1863  1 AUREOLARIA FLAVA (L.) FARW. VAR. MACRANTHA PENNELL.   SEE:
Aureolaria flava (L.) Farw. var. macrantha Pennell

1864  1 AUREOLARIA VIRGINICA (L.) FARW.   SEE:   Aureolaria
virginica (L.) Pennell

1866  1 AUREOLARIA PEDICULARIA (L.) RAF. VAR. TYPICA PENNELL.
SEE:   Aureolaria pedicularia (L.) Raf. var. pedicularia

1870  1 MELAMPYRUM LINEARE LAM. VAR. LATIFOLIUM (MUHL.) BEAUV.
SEE:   Melampyrum lineare Desr. var. latifolium Bart.

1871  1 MELAMPYRUM LINEARE LAM. VAR. PECTINATUM PENNELL.   SEE:
Melampyrum lineare Desr. var. pectinatum (Pennell) Fern.

1875  1 CAMPSIS RADICANS (L.) SEEMANN.   SEE:   Campsis radicans
(L.) Seem. ex Bureau

1877  1 CATALPA SPECIOSA WARDER.   SEE:   Catalpa speciosa (Warder
ex Barney) Engelm.

1878  1 CONOPHOLIS AMERICANA (L. F.) WALLR.   SEE:   Conopholis
americana (L.) Wallr.

1879  1 OROBANCHE LUDOVICIANA NUTT. VAR. GENUINA G. BECK.   SEE:
Orobanche ludoviciana Nutt. ssp. ludoviciana

1880  1 OROBANCHE UNIFLORA L. VAR. TYPICA ACHEY.   SEE:
Orobanche uniflora L. var. uniflora

1881   1 OROBANCHE FASCICULATA NUTT. VAR. TYPICA ACHEY.   SEE:
Orobanche fasciculata Nutt. var. fasciculata

1893   1 RUELLIA CAROLINIENSIS (WALT.) STEUD. VAR. PARVIVLORA
(NEES) BLAKE.        1893   PROBABLY NOT DISTINCT

1895   1 DIAPEDIUM BRACHIATUM (PURSH) KUNTZE.   SEE:   Dicliptera
brachiata (Pursh) Spreng.

1896   1 DIANTHERA AMERICANA L.   SEE:   Justicia americana (L.)
Vahl

1903   1 PLANTAGO PURSHII R. & S.   SEE:   Plantago patagonica
Jacq. var. patagonica

1906   1 HOUSTONIA CAERULEA L.   SEE:   Hedyotis caerulea (L.)
Hook.

1907   1 HOUSTONIA PURPUREA L.   SEE:   Hedyotis purpurea (L.)
Torr. & Gray

1908   1 HOUSTONIA ANGUSTIFOLIA MICHX.   SEE:   Hedyotis nigricans
(Lam.) Fosberg

1909   1 HOUSTONIA LONGIFOLIA GAERTN.   SEE:   Hedyotis longifolia
(Gaertn.) Hook.

1911   1 CEPHALANTHUS OCCIDENTALIS L. VAR. PUBESCENS RAF.
1911   PROBABLY NOT DISTINCT

1913   1 DIODIA TERES WALT. VAR. SETIFOLIA FERN. & GRISC.   SEE:
Diodia teres Walt.

1915   1 GALIUM CIRCAEZANS MICHX. VAR. TYPICUM FERN.   SEE:
Galium circaezans Michx. var. circaezans

1919   1 GALIUM BOREALE L. VAR. TYPICUM BECK VON MAN.   SEE:
Galium boreale L.

1920   1 GALIUM BOREALE L. VAR. INTERMEDIUM DC.        1920
PROBABLY NOT DISTINCT

1921   1 GALIUM BOREALE L. VAR. HYSSOPIFOLIUM (HOFFM.) DC.
1921   PROBABLY NOT DISTINCT

1932   1 SAMBUCUS PUBENS MICHX.   SEE:   Sambucus racemosa L. ssp.
pubens (Michx.) House

1939  1 VIBURNUM AFFINE BUSH.   SEE:   Viburnum rafinesquianum
      Schultes var. affine (Bush) House

1940  1 VIBURNUM AFFINE BUSH. VAR. HYPOMALACUM BLAKE.   SEE:
      Viburnum rafinesquianum Schultes var. rafinesquianum

1942  1 VIBURNUM PUBESCENS (AIT.) PURSH VAR. DEAMII REHD.   SEE:
      Viburnum dentatum L. var. deamii (Rehd.) Fern.

1943  1 VIBURNUM PUBESCENS (AIT.) PURSH VAR. INDIANENSE REHD.
      1943   PROBABLY NOT DISTINCT

1950  1 LINNAEA BOREALIS L. VAR. AMERICANA (FORBES) REHD.   SEE:
      Linnaea borealis L. ssp. americana (Forbes) Hulten

1951  1 LONICERA CANADENSIS MARSH.   SEE:   Lonicera canadensis
      Bartr.

1954  1 LONICERA DIOICA L. VAR. GLAUCESCENS (RYDB.) BUTTERS.
      SEE:   Lonicera dioica L. var. glaucescens (Rydb.) Butters

1958  1 VALERIANELLA INTERMEDIA DYAL.   SEE:   Valerianella
      umbilicata (Sullivant) Wood

1960  1 VALERIANA EDULIS NUTT.   SEE:   Valeriana edulis Nutt. ex
      Torr. & Gray

1961  2 DIPSACUS SYLVESTRIS HUDS.   SEE:   Dipsacus fullonum L.

1963  1 ECHINOCYSTIS LOBATA (MICHX.) T. & G.   SEE:   Echinocystis
      lobata (Michx.) Torr. & Gray

1966  1 CAMPANULA ULIGINOSA RYDB.   SEE:   Campanula aparinoides
      Pursh

1967  1 CAMPANULA APARINOIDES PURSH.        1967   PROBABLY NOT
      DISTINCT

1969  1 CAMPANULA ROTUNDIFOLIA L. VAR. INTERCEDENS (WITASEK) FARW.
      1969   PROBABLY NOT DISTINCT

1970  1 SPECULARIA PERFOLIATA (L.) A. DC.   SEE:   Triodanis
      perfoliata (L.) Nieuwl.

1971  1 LOBELIA CARDINALIS L.   SEE:   Lobelia cardinalis L. ssp.
      cardinalis

1978  1 VERNONIA ALTISSIMA NUTT.     SEE:     Vernonia gigantea
      (Walt.) Trel. ex Branner & Coville

1993  1 KUHNIA EUPATORIOIDES L.     SEE:     Brickellia eupatorioides
      (L.) Shinners

1994  1 KUHNIA EUPATORIOIDES L. VAR. CORYMBULOSA T. & G.     SEE:
      Brickellia eupatorioides (L.) Shinners var. corymbulosa
      (Torr. & Gray) Shinners

1995  1 LIATRIS SQUARROSA WILLD.     SEE:     Liatris squarrosa (L.)
      Michx.

1997  1 LIATRIS BEBBIANA RYDB.     SEE:     Liatris pycnostachya
      Michx. var. pycnostachya

2001  1 CHRYSOPSIS VILLOSA (PURSH) NUTT.     SEE:     Heterotheca
      villosa (Pursh) Shinners

2008  1 SOLIDAGO LATIFOLIA L.     SEE:     Solidago flexicaulis L.

2010  1 SOLIDAGO RACEMOSA GREENE VAR. GILLMANI (GRAY) FERN.     SEE:
      Solidago spathulata DC. ssp. spathulata var. gillmanii (Gray)
      Cronq.

2013  1 SOLIDAGO GLABERRIMA MARTENS.     SEE:     Solidago
      missouriensis Nutt. var. fasciculata Holz.

2015  1 SOLIDAGO GIGANTEA AIT. VAR. LEIOPHYLLA FERN.     SEE:
      Solidago gigantea Ait. var. serotina (Ait.) Cronq.

2016  1 SOLIDAGO ALTISSIMA L.     SEE:     Solidago canadensis L. var.
      scabra (Muhl.) Torr. & Gray

2018  1 SOLIDAGO NEMORALIS AIT. VAR. DECEMFLORA (DC.) FERN.     SEE:
      Solidago nemoralis Ait. var. longipetiolata (Mackenzie &
      Bush) Palmer & Steyermark

2021  1 SOLIDAGO RUGOSA MILL.     SEE:     Solidago rugosa Ait. var.
      rugosa

2023  1 SOLIDAGO UNILIGULATA (DC.) PORTER.     SEE:     Solidago
      uliginosa Nutt.

2029  1 SOLIDAGO GRAMINIFOLIA (L.) SALISB. VAR. NUTTALLII (GREENE)
      FERN.     SEE:     Euthamia graminifolia (L.) Nutt. ex Cass. var.
      nuttallii (Greene) W. Stone

2030  1 SOLIDAGO MEDIA (GREENE) BUSH.     SEE:     Euthamia
      gymnospermoides Greene

2031  1 SOLIDAGO REMOTA (GREENE) FRIESNER.  SEE:  Euthamia
tenuifolia (Pursh) Greene

2038  1 ASTER SAGITTIFOLIUS WEDEMEYER EX WILLD.  SEE:  Aster X
sagittifolius Wedemeyer ex Willd.

2039  1 ASTER SAGITTIFOLIUS WEDEMEYER EX WILLD. VAR. UROPHYLLUS
LINDL.  SEE:  Aster urophyllus Lindl.

2045  1 ASTER OBLONGIFOLIUS NUTT. VAR. RIGIDULUS GRAY.
2045  PROBABLY NOT DISTINCT

2049  1 ASTER LUCIDULUS (GRAY) WIEG.  SEE:  Aster firmus Nees

2052  1 ASTER JUNCEUS AIT.  SEE:  Aster borealis (Torr. & Gray)
Prov.

2053  1 ASTER PANICULATUS LAM.  SEE:  Aster simplex Willd. var.
ramosissimus (Torr. & Gray) Cronq.

2054  1 ASTER PANICULATUS LAM. VAR. SIMPLEX (WILLD.) BURGESS.
SEE:  Aster simplex Willd. var. simplex

2055  1 ASTER INTERIOR WIEG.  SEE:  Aster X interior Wieg.

2058  1 ASTER EXIGUUS (FERN.) RYDB.  SEE:  Aster ericoides L.
ssp. ericoides var. ericoides

2059  1 ASTER ERICOIDES L.        2059  PROBABLY NOT DISTINCT

2062  1 ASTER MISSOURIENSIS BRITTON.  SEE:  Aster ontarionis
Wieg.

2064  1 ASTER PTARMICOIDES (NEES) T. & G.  SEE:  Solidago
ptarmicoides (Nees) Boivin

2069  1 ERIGERON CANADENSIS L.  SEE:  Conyza canadensis (L.)
Cronq.

2070  1 ERIGERON PUSILLUS NUTT.  SEE:  Conyza canadensis (L.)
Cronq. var. pusilla (Nutt.) Cronq.

2071  1 ERIGERON DIVARICATUS MICHX.  SEE:  Conyza ramosissima
Cronq.

2074 1 ERIGERON RAMOSUS (WALT.) BSP.     SEE:     Erigeron strigosus
Muhl. ex Willd. var. strigosus

2076 1 SERICOCARPUS LINIFOLIUS (L.) BSP.     SEE:     Aster
solidagineus Michx.

2077 1 PLUCHEA VISCIDA (RAF.) HOUSE.     SEE:     Pluchea camphorata
(L.) DC.

2079 1 ANTENNARIA NEODIOICA GREENE.     SEE:     Antennaria neglecta
Greene var. attenuata (Fern.) Cronq.

2081 1 ANTENNARIA PARLINII FERN.     SEE:     Antennaria
plantaginifolia (L.) Richards var. arnoglossa (Greene) Cronq.

2083 1 ANTENNARIA FALLAX GREENE.     SEE:     Antennaria
plantaginifolia (L.) Richards var. ambigens (Greene) Cronq.

2084 1 ANTENNARIA FALLAX GREENE VAR. CALOPHYLLA (GREENE) FERN.
2084    PROBABLY NOT DISTINCT

2085 1 ANTENNARIA MUNDA FERN.          2085    PROBABLY NOT
DISTINCT

2086 1 ANAPHALIS MARGARITACEA (L.) GRAY VAR. REVOLUTA SUKSD.
FORMA ARACHNOIDEA FERN.          2086    PROBABLY NOT DISTINCT

2088 1 GNAPHALIUM MACOUNII GREENE.     SEE:     Gnaphalium viscosum
H.B.K.

2096A 1 SILPHIUM LACINIATUM L. VAR. ROBINSONII PERRY.     SEE:
Silphium laciniatum L. var. robinsonii Perry

2097 1 SILPHIUM TRIFOLIATUM L.     SEE:     Silphium asteriscus L.
ssp. trifoliatum (Ell.) Weber & T.R. Fisher ined.

2099 1 SILPHIUM INTEGRIFOLIUM L. VAR. DEAMII PERRY.     SEE:
Silphium integrifolium Michx. var. deamii Perry

2101 1 IVA CILIATA WILLD.     SEE:     Iva annua L. var. annua

2105 1 AMBROSIA ELATIOR L.     SEE:     Ambrosia artemisiifolia L.
var. elatior Descourtils

2108 1 XANTHIUM PENNSYLVANICUM WALLR.     SEE:     Xanthium
strumarium L. var. canadense (P. Mill.) Torr. & Gray

2109 1 XANTHIUM ITALICUM MORETTI.          2109    PROBABLY NOT
DISTINCT

2111 1 ECLIPTA ALBA (L.) HASSK.   SEE:   Eclipta prostrata (L.)
L.

2116 1 RUDBECKIA FULGIDA AIT.    SEE:    Rudbeckia fulgida Ait.
var. fulgida

2117 1 RUDBECKIA UMBROSA BOYNTON & BEADLE.   SEE:   Rudbeckia
fulgida Ait. var. umbrosa (C.L. Boynt. & Beadle)   Cronq.

2118 1 RUDBECKIA SULLIVANTII BOYNTON & BEADLE.   SEE:   Rudbeckia
fulgida Ait. var. sullivantii (C.L. Boynt. & Beadle)   Cronq.

2119 1 RUDBECKIA PALUSTRIS EGGERT.   SEE:   Rudbeckia fulgida
Ait. var. palustris (Eggert) Perdue

2120 1 RUDBECKIA DEAMII BLAKE.   SEE:   Rudbeckia fulgida Ait.
var. deamii (Blake) Perdue

2121 1 BRAUNERIA PURPUREA (L.) BRITT.   SEE:   Echinacea purpurea
(L.) Moench

2122 1 BRAUNERIA PALLIDA (NUTT.) BRITT.   SEE:   Echinacea
pallida Nutt.

2133 1 HELIANTHUS DORONICOIDES LAM.   SEE:   Helianthus X
doronicoides Lam.

2141 1 ACTINOMERIS ALTERNIFOLIA (L.) DC.   SEE:   Verbesina
alternifolia (L.) Britt.

2147 1 COREOPSIS TRIPTERIS L. VAR. DEAMII STANDLEY.          2147
PROBABLY NOT DISTINCT

2149 1 BIDENS COMOSA (GRAY) WIEG.   SEE:   Bidens tripartita L.

2156 1 BIDENS ARISTOSA (MICHX.) BRITT. VAR. FRITCHEYI FERN.
2156   PROBABLY NOT DISTINCT

2157 1 BIDENS ARISTOSA (MICHX.) BRITT. VAR. MUTICA GRAY EX
GATTINGER.          2157   PROBABLY NOT DISTINCT

2160 1 MEGALODONTA BECKII (TORR.) GREENE.   SEE:   Megalodonta
beckii (Torr. ex Spreng.) Greene

2161  2 GALINSOGA CILIATA (RAF.) BLAKE.   SEE:   Galinsoga quadriradiata Ruiz & Pavon

2162  2 HYMENOPAPPUS CAROLINENSIS (LAM.) PORTER.   SEE: Hymenopappus scabiosaeus L'Her.

2163  2 HELENIUM TENUIFOLIUM NUTT.   SEE:   Helenium amarum (Raf.) H. Rock

2165  1 HELENIUM NUDIFLORUM NUTT.   SEE:   Helenium flexuosum Raf.

2166  2 DYSSODIA PAPPOSA (VENT.) HITCHC.   SEE:   Dyssodia papposa (Vent.) A.S. Hitchc.

2169  2 ANTHEMIS NOBILIS L.   SEE:   Chamaemelum nobilis (L.) All.

2172  2 CHRYSANTHEMUM LEUCANTHEMUM L. VAR. PINNATIFIDUM LECOQ. & LAMOTTE.   SEE:   Leucanthemum vulgare Lam.

2173  2 CHRYSANTHEMUM BALSAMITA L. VAR. TANACETOIDES BOISS.   SEE: Balsamita major Desf.

2178  1 ARTEMISIA CAUDATA MICHX.   SEE:   Artemisia campestris L. ssp. caudata (Michx.) Hall & Clements

2179  2 ARTEMISIA GNAPHALODES NUTT.   SEE:   Artemisia ludoviciana Nutt. ssp. ludoviciana

2180  1 ERECHTITES HIERACIFOLIA (L.) RAF.   SEE:   Erechtites hieraciifolia (L.) Raf. ex DC.

2188  1 SENECIO OBOVATUS MUHL.   SEE:   Senecio obovatus Muhl. ex Willd.

2191  1 SENECIO PAUPERCULUS MICHX. VAR. BALSAMITAE (MUHL.) FERN. SEE:   Senecio pauperculus Michx.

2192  2 ARCTIUM MINUS (HILL) BERNH.   SEE:   Arctium minus Bernh.

2204  1 SERINIA OPPOSITIFOLIA (RAF.) KTZE.   SEE:   Krigia caespitosa (Raf.) Chambers

2209  2 TRAGOPOGON PRATENSIS L.   SEE:   Tragopogon pratensis L. ssp. pratensis

2210  2 TARAXACUM PALUSTRE (LYONS) LAM. & DC. VAR. VULGARE (LAM.)
      FERN.   SEE:   Taraxacum palustre (Lyons) Symons

2212  2 SONCHUS ARVENSIS L.   SEE:   Sonchus arvensis L. ssp.
      arvensis

2213  2 SONCHUS ARVENSIS L. VAR. GLABRESCENS GUENTHER, GRAB &
      WIMM.           2213   PROBABLY NOT DISTINCT

2217  2 LACTUCA SCARIOLA L. VAR. INTEGRATA GREN. & GODR.   SEE:
      Lactuca sativa L.

2218  1 LACTUCA CAMPESTRIS GREENE.   SEE:   Lactuca ludoviciana
      (Nutt.) Riddell

2220  1 LACTUCA CANADENSIS L. VAR. TYPICA WIEG.   SEE:   Lactuca
      canadensis L. var. canadensis

2224  1 LACTUCA VILLOSA JACQ.   SEE:   Lactuca floridana (L.)
      Gaertn. var. villosa (Jacq.) Cronq.

2226  1 LACTUCA SPICATA (LAM.) HITCHC.   SEE:   Lactuca biennis
      (Moench) Fern.

2227  1 LACTUCA SPICATA (LAM) HITCHC. VAR. INTEGRIFOLIA (GRAY)
      BRITT.           2227   PROBABLY NOT DISTINCT

# Vascular Plant Families Reported from Indiana

| | | N. Amer. Genera | Indiana Genera | N. Amer. Species | Indiana Species |
|---|---|---|---|---|---|
| 1 | ADIANTACEAE | 21 | 3 | 136 | 4 |
| 1 | ASPLENIACEAE | 32 | 11 | 265 | 26 |
| 2 | AZOLLACEAE | 1 | 1 | 3 | 1 |
| 2 | BLECHNACEAE | 5 | 1 | 22 | 1 |
| 2 | DENNSTAEDTIACEAE | 10 | 2 | 26 | 2 |
| 2 | EQUISETACEAE | 1 | 1 | 11 | 5 |
| 3 | HYMENOPHYLLACEAE | 2 | 1 | 46 | 1 |
| 3 | ISOETACEAE | 1 | 1 | 23 | 1 |
| 3 | LYCOPODIACEAE | 1 | 1 | 40 | 7 |
| 3 | OPHIOGLOSSACEAE | 2 | 2 | 28 | 7 |
| 3 | OSMUNDACEAE | 1 | 1 | 3 | 3 |
| 3 | POLYPODIACEAE | 2 | 1 | 36 | 2 |
| 4 | SELAGINELLACEAE | 1 | 1 | 50 | 2 |
| 4 | CUPRESSACEAE | 6 | 2 | 30 | 3 |
| 4 | PINACEAE | 6 | 3 | 68 | 5 |
| 4 | TAXACEAE | 2 | 1 | 5 | 1 |
| 4 | TAXODIACEAE | 4 | 1 | 6 | 1 |
| 4 | ACANTHACEAE | 24 | 3 | 98 | 5 |
| 4 | ACERACEAE | 1 | 1 | 19 | 6 |
| 5 | AGAVACEAE | 11 | 2 | 96 | 3 |
| 5 | AIZOACEAE | 16 | 1 | 29 | 1 |
| 5 | ALISMACEAE | 4 | 3 | 35 | 9 |
| 5 | AMARANTHACEAE | 18 | 3 | 108 | 12 |
| 6 | ANACARDIACEAE | 11 | 2 | 34 | 6 |
| 6 | ANNONACEAE | 6 | 1 | 20 | 1 |
| 6 | APIACEAE | 83 | 28 | 391 | 37 |
| 7 | APOCYNACEAE | 27 | 4 | 78 | 5 |
| 7 | AQUIFOLIACEAE | 2 | 2 | 31 | 3 |
| 8 | ARACEAE | 21 | 5 | 41 | 6 |
| 8 | ARALIACEAE | 11 | 2 | 49 | 6 |
| 8 | ARISTOLOCHIACEAE | 3 | 2 | 35 | 3 |
| 8 | ASCLEPIADACEAE | 15 | 3 | 145 | 17 |
| 9 | ASTERACEAE | 346 | 72 | 2687 | 255 |
| 16 | BALSAMINACEAE | 1 | 1 | 10 | 2 |
| 16 | BERBERIDACEAE | 10 | 4 | 32 | 5 |
| 16 | BETULACEAE | 5 | 5 | 33 | 11 |
| 17 | BIGNONIACEAE | 18 | 4 | 28 | 5 |
| 17 | BORAGINACEAE | 34 | 11 | 384 | 21 |
| 17 | BRASSICACEAE | 94 | 36 | 634 | 72 |
| 19 | CACTACEAE | 19 | 1 | 174 | 1 |
| 19 | CALLITRICHACEAE | 1 | 1 | 13 | 2 |
| 19 | CAMPANULACEAE | 23 | 3 | 290 | 11 |
| 20 | CAPPARIDACEAE | 9 | 1 | 43 | 1 |
| 20 | CAPRIFOLIACEAE | 8 | 7 | 80 | 27 |
| 21 | CARYOPHYLLACEAE | 35 | 18 | 326 | 43 |
| 22 | CELASTRACEAE | 13 | 2 | 32 | 5 |
| 22 | CERATOPHYLLACEAE | 1 | 1 | 3 | 1 |
| 22 | CHENOPODIACEAE | 25 | 7 | 187 | 23 |
| 23 | CISTACEAE | 4 | 3 | 35 | 9 |
| 23 | CLUSIACEAE | 6 | 2 | 68 | 21 |
| 24 | COMMELINACEAE | 13 | 2 | 49 | 7 |
| 24 | CONVOLVULACEAE | 18 | 4 | 198 | 18 |
| 25 | CORNACEAE | 1 | 1 | 14 | 9 |
| 25 | CRASSULACEAE | 12 | 1 | 108 | 4 |

| | | N. Amer. Genera | Indiana Genera | N. Amer. Species | Indiana Species |
|---|---|---|---|---|---|
| 25 | CUCURBITACEAE | 26 | 5 | 76 | 5 |
| 25 | CYPERACEAE | 26 | 14 | 959 | 222 |
| 31 | DIOSCOREACEAE | 2 | 1 | 13 | 4 |
| 31 | DIPSACACEAE | 6 | 1 | 12 | 2 |
| 31 | DROSERACEAE | 2 | 1 | 8 | 2 |
| 32 | EBENACEAE | 1 | 1 | 7 | 1 |
| 32 | ELAEAGNACEAE | 3 | 1 | 10 | 1 |
| 32 | ERICACEAE | 44 | 13 | 219 | 25 |
| 32 | ERIOCAULACEAE | 3 | 1 | 16 | 1 |
| 32 | EUPHORBIACEAE | 47 | 8 | 358 | 27 |
| 33 | FABACEAE | 142 | 32 | 1521 | 93 |
| 36 | FAGACEAE | 5 | 3 | 88 | 20 |
| 36 | GENTIANACEAE | 17 | 8 | 117 | 15 |
| 37 | GERANIACEAE | 3 | 2 | 62 | 7 |
| 37 | HALORAGIDACEAE | 5 | 2 | 24 | 5 |
| 37 | HAMAMELIDACEAE | 3 | 2 | 5 | 2 |
| 37 | HIPPOCASTANACEAE | 1 | 1 | 7 | 3 |
| 38 | HIPPURIDACEAE | 1 | 1 | 3 | 1 |
| 38 | HYDROCHARITACEAE | 10 | 1 | 24 | 3 |
| 38 | HYDROPHYLLACEAE | 17 | 3 | 238 | 8 |
| 38 | IRIDACEAE | 18 | 3 | 109 | 9 |
| 38 | JUGLANDACEAE | 2 | 2 | 21 | 11 |
| 39 | JUNCACEAE | 2 | 2 | 130 | 24 |
| 40 | LAMIACEAE | 70 | 30 | 482 | 70 |
| 42 | LARDIZABALACEAE | 1 | 1 | 1 | 1 |
| 42 | LAURACEAE | 13 | 2 | 35 | 2 |
| 42 | LEMNACEAE | 4 | 4 | 17 | 12 |
| 42 | LENTIBULARIACEAE | 2 | 1 | 31 | 9 |
| 43 | LILIACEAE | 76 | 26 | 487 | 49 |
| 44 | LIMNANTHACEAE | 2 | 1 | 10 | 1 |
| 44 | LINACEAE | 4 | 1 | 48 | 6 |
| 44 | LOGANIACEAE | 8 | 1 | 47 | 1 |
| 44 | LORANTHACEAE | 8 | 1 | 49 | 1 |
| 44 | LYTHRACEAE | 11 | 6 | 36 | 7 |
| 45 | MAGNOLIACEAE | 2 | 2 | 11 | 3 |
| 45 | MALVACEAE | 41 | 7 | 266 | 14 |
| 45 | MARTYNIACEAE | 4 | 1 | 9 | 1 |
| 45 | MELASTOMATACEAE | 19 | 1 | 65 | 2 |
| 45 | MENISPERMACEAE | 5 | 3 | 11 | 3 |
| 45 | MENYANTHACEAE | 3 | 1 | 6 | 1 |
| 46 | MORACEAE | 17 | 4 | 31 | 6 |
| 46 | MYRICACEAE | 2 | 1 | 11 | 1 |
| 46 | NAJADACEAE | 1 | 1 | 8 | 4 |
| 46 | NYCTAGINACEAE | 16 | 1 | 119 | 4 |
| 46 | NYMPHAEACEAE | 5 | 5 | 18 | 5 |
| 46 | NYSSACEAE | 1 | 1 | 3 | 1 |
| 47 | OLEACEAE | 12 | 4 | 65 | 8 |
| 47 | ONAGRACEAE | 13 | 6 | 252 | 30 |
| 48 | ORCHIDACEAE | 88 | 17 | 285 | 41 |
| 49 | OROBANCHACEAE | 4 | 3 | 23 | 5 |
| 49 | OXALIDACEAE | 1 | 1 | 30 | 6 |
| 49 | PAPAVERACEAE | 23 | 8 | 93 | 10 |
| 50 | PASSIFLORACEAE | 1 | 1 | 26 | 2 |
| 50 | PHYTOLACCACEAE | 8 | 1 | 14 | 1 |

| | | N. Amer. Genera | Indiana Genera | N. Amer. Species | Indiana Species |
|----|------------------|------|------|------|------|
| 50 | PLANTAGINACEAE | 2 | 1 | 43 | 9 |
| 50 | PLATANACEAE | 1 | 1 | 4 | 1 |
| 50 | POACEAE | 231 | 75 | 1490 | 223 |
| 56 | POLEMONIACEAE | 14 | 4 | 283 | 13 |
| 56 | POLYGALACEAE | 3 | 1 | 64 | 6 |
| 57 | POLYGONACEAE | 24 | 4 | 446 | 34 |
| 58 | PONTEDERIACEAE | 5 | 2 | 11 | 3 |
| 58 | PORTULACACEAE | 8 | 3 | 108 | 3 |
| 58 | POTAMOGETONACEAE | 3 | 1 | 43 | 19 |
| 59 | PRIMULACEAE | 10 | 7 | 90 | 17 |
| 59 | RANUNCULACEAE | 24 | 17 | 323 | 45 |
| 60 | RHAMNACEAE | 14 | 2 | 123 | 7 |
| 61 | ROSACEAE | 62 | 21 | 870 | 101 |
| 63 | RUBIACEAE | 60 | 6 | 317 | 23 |
| 64 | RUTACEAE | 20 | 2 | 130 | 2 |
| 64 | SALICACEAE | 2 | 2 | 117 | 26 |
| 65 | SANTALACEAE | 8 | 1 | 16 | 1 |
| 65 | SAPINDACEAE | 17 | 1 | 38 | 1 |
| 65 | SAPOTACEAE | 9 | 1 | 39 | 1 |
| 65 | SARRACENIACEAE | 2 | 1 | 9 | 1 |
| 65 | SAURURACEAE | 2 | 1 | 2 | 1 |
| 65 | SAXIFRAGACEAE | 36 | 11 | 285 | 19 |
| 66 | SCHEUCHZERIACEAE | 3 | 2 | 8 | 3 |
| 66 | SCROPHULARIACEAE | 69 | 24 | 838 | 59 |
| 68 | SIMAROUBACEAE | 9 | 1 | 13 | 1 |
| 68 | SMILACACEAE | 1 | 1 | 25 | 8 |
| 68 | SOLANACEAE | 33 | 7 | 199 | 15 |
| 69 | SPARGANIACEAE | 1 | 1 | 9 | 4 |
| 69 | STAPHYLEACEAE | 2 | 1 | 3 | 1 |
| 69 | STYRACACEAE | 2 | 1 | 10 | 1 |
| 69 | THYMELIACEAE | 6 | 1 | 36 | 1 |
| 69 | TILIACEAE | 3 | 1 | 16 | 2 |
| 69 | TYPHACEAE | 1 | 1 | 4 | 2 |
| 69 | ULMACEAE | 5 | 2 | 24 | 8 |
| 70 | URTICACEAE | 13 | 5 | 72 | 6 |
| 70 | VALERIANACEAE | 4 | 2 | 35 | 7 |
| 70 | VERBENACEAE | 23 | 4 | 135 | 10 |
| 71 | VIOLACEAE | 4 | 2 | 107 | 24 |
| 72 | VITACEAE | 4 | 3 | 40 | 10 |
| 72 | XYRIDACEAE | 1 | 1 | 22 | 2 |
| 72 | ZANNICHELLIACEAE | 3 | 1 | 3 | 1 |
| 72 | ZYGOPHYLLACEAE | 7 | 1 | 20 | 1 |

# Alphabetical Index of Families and
# Genera Reported from Indiana

| | | | | | | |
|---|---|---:|---|---|---|---:|
| 18 | BARBAREA | 122 | 20 | CAPRIFOLIACEAE | 147 |
| 36 | BARTONIA | 242 | 18 | CAPSELLA | 123 |
| 38 | BELAMCANDA | 254 | 18 | CARDAMINE | 123 |
| 10 | BELLIS | 59 | 18 | CARDARIA | 124 |
| 16 | BERBERIDACEAE | 110 | 11 | CARDUUS | 62 |
| 16 | BERBERIS | 110 | 25 | CAREX | 173 |
| 18 | BERTEROA | 122 | 16 | CARPINUS | 112 |
| 67 | BESSEYA | 440 | 6 | CARUM | 32 |
| 16 | BETULA | 111 | 38 | CARYA | 256 |
| 16 | BETULACEAE | 111 | 21 | CARYOPHYLLACEAE | 149 |
| 10 | BIDENS | 59 | 33 | CASSIA | 214 |
| 17 | BIGNONIA | 112 | 36 | CASTANEA | 239 |
| 17 | BIGNONIACEAE | 112 | 67 | CASTILLEJA | 440 |
| 2 | BLECHNACEAE | 9 | 17 | CATALPA | 112 |
| 40 | BLEPHILIA | 261 | 16 | CAULOPHYLLUM | 110 |
| 70 | BOEHMERIA | 463 | 60 | CEANOTHUS | 388 |
| 11 | BOLTONIA | 60 | 22 | CELASTRACEAE | 157 |
| 17 | BORAGINACEAE | 113 | 22 | CELASTRUS | 157 |
| 3 | BOTRYCHIUM | 15 | 69 | CELTIS | 463 |
| 51 | BOUTELOUA | 330 | 51 | CENCHRUS | 333 |
| 51 | BRACHYELYTRUM | 330 | 11 | CENTAUREA | 62 |
| 46 | BRASENIA | 299 | 36 | CENTAURIUM | 242 |
| 18 | BRASSICA | 122 | 63 | CEPHALANTHUS | 411 |
| 17 | BRASSICACEAE | 120 | 21 | CERASTIUM | 150 |
| 11 | BRICKELLIA | 61 | 22 | CERATOPHYLLACEAE | 157 |
| 51 | BROMUS | 330 | 22 | CERATOPHYLLUM | 157 |
| 67 | BUCHNERA | 440 | 34 | CERC&S | 215 |
| 17 | BUGLOSSOIDES | 114 | 67 | CHAENORRHINUM | 442 |
| 25 | BULBOSTYLIS | 173 | 6 | CHAEROPHYLLUM | 32 |
| 65 | BUMELIA | 430 | 32 | CHAMAEDAPHNE | 192 |
| 6 | BUPLEURUM | 32 | 43 | CHAMAELIRIUM | 274 |
| 46 | CABOMBA | 299 | 11 | CHAMAEMELUM | 63 |
| 11 | CACALIA | 61 | 33 | CHAMAESYCE | 198 |
| 19 | CACTACEAE | 136 | 51 | CHASMANTHIUM | 333 |
| 18 | CAKILE | 123 | 1 | CHEILANTHES | 1 |
| 51 | CALAMAGROSTIS | 332 | 49 | CHELIDONIUM | 319 |
| 40 | CALAMINTHA | 261 | 67 | CHELONE | 442 |
| 51 | CALAMOVILFA | 332 | 22 | CHENOPODIACEAE | 158 |
| 8 | CALLA | 40 | 22 | CHENOPODIUM | 159 |
| 45 | CALLIRHOE | 287 | 32 | CHIMAPHILA | 192 |
| 19 | CALLITRICHACEAE | 141 | 51 | CHLORIS | 333 |
| 19 | CALLITRICHE | 141 | 18 | CHORISPORA | 124 |
| 48 | CALOPOGON | 310 | 65 | CHRYSOSPLENIUM | 431 |
| 59 | CALTHA | 382 | 11 | CICHORIUM | 65 |
| 45 | CALYCOCARPUM | 293 | 6 | CICUTA | 32 |
| 47 | CALYLOPHUS | 302 | 59 | CIMICIFUGA | 382 |
| 24 | CALYSTEGIA | 165 | 51 | CINNA | 333 |
| 43 | CAMASSIA | 274 | 47 | CIRCAEA | 304 |
| 18 | CAMELINA | 123 | 11 | CIRSIUM | 65 |
| 19 | CAMPANULA | 142 | 23 | CISTACEAE | 161 |
| 19 | CAMPANULACEAE | 141 | 25 | CITRULLUS | 172 |
| 17 | CAMPSIS | 112 | 29 | CLADIUM | 181 |
| 46 | CANNABIS | 294 | 33 | CLADRASTIS | 215 |
| 20 | CAPPARIDACEAE | 146 | 58 | CLAYTONIA | 375 |

129

| | | | | | | |
|---|---|---:|---|---|---|---:|
| 59 | CLEMATIS | 382 | | 1 | CYSTOPTERIS | 4 |
| 40 | CLINOPODIUM | 261 | | 51 | DACTYLIS | 334 |
| 43 | CLINTONIA | 274 | | 34 | DALEA | 216 |
| 33 | CLITORIA | 215 | | 52 | DANTHONIA | 334 |
| 23 | CLUSIACEAE | 162 | | 67 | DASYSTOMA | 443 |
| 45 | COCCULUS | 293 | | 68 | DATURA | 456 |
| 48 | COELOGLOSSUM | 310 | | 6 | DAUCUS | 33 |
| 67 | COLLINSIA | 442 | | 44 | DECODON | 285 |
| 40 | COLLINSONIA | 261 | | 59 | DELPHINIUM | 383 |
| 56 | COLLOMIA | 357 | | 2 | DENNSTAEDTIA | 10 |
| 65 | COMANDRA | 428 | | 2 | DENNSTAEDTIACEAE | 10 |
| 24 | COMMELINA | 164 | | 18 | DENTARIA | 124 |
| 24 | COMMELINACEAE | 163 | | 1 | DEPARIA | 5 |
| 46 | COMPTONIA | 294 | | 52 | DESCHAMPSIA | 334 |
| 6 | CONIOSELINUM | 32 | | 18 | DESCURAINIA | 125 |
| 6 | CONIUM | 32 | | 34 | DESMANTHUS | 217 |
| 49 | CONOPHOLIS | 317 | | 34 | DESMODIUM | 217 |
| 18 | CONRINGIA | 124 | | 21 | DIANTHUS | 150 |
| 59 | CONSOLIDA | 383 | | 52 | DIARRHENA | 334 |
| 43 | CONVALLARIA | 274 | | 49 | DICENTRA | 319 |
| 24 | CONVOLVULACEAE | 165 | | 52 | DICHANTHELIUM | 334 |
| 24 | CONVOLVULUS | 166 | | 4 | DICLIPTERA | 21 |
| 11 | CONYZA | 67 | | 44 | DIDIPLIS | 285 |
| 59 | COPTIS | 383 | | 20 | DIERVILLA | 147 |
| 48 | CORALLORHIZA | 311 | | 52 | DIGITARIA | 337 |
| 11 | COREOPSIS | 67 | | 63 | DIODIA | 412 |
| 23 | CORISPERMUM | 160 | | 31 | DIOSCOREA | 189 |
| 25 | CORNACEAE | 169 | | 31 | DIOSCOREACEAE | 189 |
| 25 | CORNUS | 169 | | 32 | DIOSPYROS | 190 |
| 34 | CORONILLA | 215 | | 53 | DIPLACHNE | 338 |
| 18 | CORONOPUS | 124 | | 1 | DIPLAZIUM | 5 |
| 49 | CORYDALIS | 319 | | 18 | DIPLOTAXIS | 125 |
| 16 | CORYLUS | 112 | | 31 | DIPSACACEAE | 189 |
| 25 | CRASSULACEAE | 169 | | 31 | DIPSACUS | 189 |
| 61 | CRATAEGUS | 393 | | 69 | DIRCA | 462 |
| 11 | CREPIS | 68 | | 59 | DODECATHEON | 379 |
| 34 | CROTALARIA | 215 | | 18 | DRABA | 125 |
| 33 | CROTON | 201 | | 40 | DRACOCEPHALUM | 261 |
| 33 | CROTONOPSIS | 201 | | 31 | DROSERA | 189 |
| 51 | CRYPSIS | 333 | | 31 | DROSERACEAE | 189 |
| 6 | CRYPTOTAENIA | 32 | | 1 | DRYOPTERIS | 5 |
| 25 | CUCURBITA | 172 | | 61 | DUCHESNEA | 397 |
| 25 | CUCURBITACEAE | 172 | | 30 | DULICHIUM | 183 |
| 40 | CUNILA | 261 | | 11 | DYSSODIA | 70 |
| 44 | CUPHEA | 285 | | 32 | EBENACEAE | 190 |
| 4 | CUPRESSACEAE | 18 | | 12 | ECHINACEA | 71 |
| 24 | CUSCUTA | 166 | | 53 | ECHINOCHLOA | 338 |
| 23 | CYCLOLOMA | 160 | | 25 | ECHINOCYSTIS | 172 |
| 9 | CYNANCHUM | 45 | | 5 | ECHINODORUS | 26 |
| 51 | CYNODON | 334 | | 17 | ECHIUM | 116 |
| 17 | CYNOGLOSSUM | 116 | | 12 | ECLIPTA | 71 |
| 25 | CYPERACEAE | 173 | | 32 | ELAEAGNACEAE | 190 |
| 29 | CYPERUS | 181 | | 30 | ELEOCHARIS | 183 |
| 48 | CYPRIPEDIUM | 311 | | 12 | ELEPHANTOPUS | 71 |

130

| 53 | ELEUSINE | 338 |
| 38 | ELLISIA | 250 |
| 38 | ELODEA | 250 |
| 53 | ELYMUS | 339 |
| 49 | EPIFAGUS | 317 |
| 32 | EPIGAEA | 193 |
| 47 | EPILOBIUM | 305 |
| 48 | EPIPACTIS | 312 |
| 2 | EQUISETACEAE | 10 |
| 2 | EQUISETUM | 10 |
| 53 | ERAGROSTIS | 340 |
| 12 | ERECHTITES | 72 |
| 53 | ERIANTHUS | 340 |
| 32 | ERICACEAE | 191 |
| 6 | ERIGENIA | 33 |
| 12 | ERIGERON | 72 |
| 32 | ERIOCAULACEAE | 197 |
| 32 | ERIOCAULON | 197 |
| 30 | ERIOPHORUM | 184 |
| 37 | ERODIUM | 245 |
| 18 | EROPHILA | 127 |
| 18 | ERUCASTRUM | 127 |
| 6 | ERYNGIUM | 33 |
| 18 | ERYSIMUM | 127 |
| 43 | ERYTHRONIUM | 275 |
| 22 | EUONYMUS | 157 |
| 12 | EUPATORIUM | 75 |
| 33 | EUPHORBIA | 202 |
| 32 | EUPHORBIACEAE | 197 |
| 12 | EUTHAMIA | 76 |
| 33 | FABACEAE | 205 |
| 36 | FAGACEAE | 239 |
| 57 | FAGOPYRUM | 371 |
| 36 | FAGUS | 239 |
| 53 | FESTUCA | 341 |
| 61 | FILIPENDULA | 398 |
| 30 | FIMBRISTYLIS | 185 |
| 44 | FLOERKEA | 280 |
| 47 | FORESTIERA | 300 |
| 61 | FRAGARIA | 398 |
| 36 | FRASERA | 243 |
| 47 | FRAXINUS | 301 |
| 6 | FROELICHIA | 29 |
| 30 | FUIRENA | 185 |
| 49 | FUMARIA | 320 |
| 12 | GAILLARDIA | 77 |
| 48 | GALEARIS | 312 |
| 12 | GALINSOGA | 77 |
| 63 | GALIUM | 412 |
| 32 | GAULTHERIA | 193 |
| 47 | GAURA | 306 |
| 32 | GAYLUSSACIA | 193 |
| 36 | GENTIANA | 243 |
| 36 | GENTIANACEAE | 242 |

| 37 | GENTIANELLA | 244 |
| 37 | GENTIANOPSIS | 244 |
| 37 | GERANIACEAE | 245 |
| 61 | GEUM | 398 |
| 37 | GERANIUM | 246 |
| 70 | GLANDULARIA | 466 |
| 40 | GLECHOMA | 262 |
| 34 | GLEDITSIA | 219 |
| 53 | GLYCERIA | 342 |
| 34 | GLYCYRRHIZA | 219 |
| 12 | GNAPHALIUM | 78 |
| 48 | GOODYERA | 312 |
| 67 | GRATIOLA | 444 |
| 12 | GRINDELIA | 78 |
| 34 | GYMNOCLADUS | 220 |
| 53 | GYMNOPOGON | 342 |
| 21 | GYPSOPHILA | 150 |
| 17 | HACKELIA | 116 |
| 37 | HALORAGIDACEAE | 248 |
| 37 | HAMAMELIDACEAE | 249 |
| 37 | HAMAMELIS | 249 |
| 40 | HEDEOMA | 262 |
| 64 | HEDYOTIS | 415 |
| 12 | HELENIUM | 80 |
| 23 | HELIANTHEMUM | 161 |
| 12 | HELIANTHUS | 81 |
| 13 | HELIOPSIS | 82 |
| 17 | HELIOTROPIUM | 116 |
| 43 | HEMEROCALLIS | 276 |
| 30 | HEMICARPHA | 185 |
| 59 | HEPATICA | 384 |
| 6 | HERACLEUM | 33 |
| 19 | HESPERIS | 128 |
| 58 | HETERANTHERA | 375 |
| 13 | HETEROTHECA | 83 |
| 65 | HEUCHERA | 432 |
| 48 | HEXALECTRIS | 312 |
| 45 | HIBISCUS | 287 |
| 13 | HIERACIUM | 84 |
| 53 | HIEROCHLOE | 342 |
| 37 | HIPPOCASTANACEAE | 249 |
| 38 | HIPPURIDACEAE | 249 |
| 38 | HIPPURIS | 249 |
| 53 | HOLCUS | 343 |
| 21 | HOLOSTEUM | 151 |
| 53 | HORDEUM | 343 |
| 59 | HOTTONIA | 380 |
| 23 | HUDSONIA | 162 |
| 46 | HUMULUS | 294 |
| 71 | HYBANTHUS | 468 |
| 66 | HYDRANGEA | 433 |
| 59 | HYDRASTIS | 385 |
| 38 | HYDROCHARITACEAE | 249 |
| 6 | HYDROCOTYLE | 34 |

| | | |
|---|---|---:|
| 38 | HYDROPHYLLACEAE | 250 |
| 38 | HYDROPHYLLUM | 250 |
| 43 | HYMENOCALLIS | 276 |
| 13 | HYMENOPAPPUS | 85 |
| 7 | HYMENOPHYLLACEAE | 11 |
| 23 | HYPERICUM | 162 |
| 13 | HYPOCHOERIS | 86 |
| 43 | HYPOXIS | 276 |
| 53 | HYSTRIX | 343 |
| 7 | ILEX | 40 |
| 16 | IMPATIENS | 110 |
| 13 | INULA | 86 |
| 19 | IODANTHUS | 128 |
| 24 | IPOMOEA | 167 |
| 56 | IPOMOPSIS | 358 |
| 6 | IRESINE | 29 |
| 38 | IRIDACEAE | 254 |
| 38 | IRIS | 254 |
| 3 | ISOETACEAE | 12 |
| 3 | ISOETES | 12 |
| 60 | ISOPYRUM | 385 |
| 48 | ISOTRIA | 313 |
| 13 | IVA | 86 |
| 16 | JEFFERSONIA | 110 |
| 38 | JUGLANDACEAE | 256 |
| 39 | JUGLANS | 257 |
| 39 | JUNCACEAE | 257 |
| 39 | JUNCUS | 257 |
| 4 | JUNIPERUS | 18 |
| 4 | JUSTICIA | 21 |
| 32 | KALMIA | 193 |
| 67 | KICKXIA | 444 |
| 23 | KOCHIA | 160 |
| 53 | KOELERIA | 343 |
| 65 | KOELREUTERIA | 429 |
| 13 | KRIGIA | 86 |
| 34 | KUMMEROWIA | 220 |
| 13 | LACTUCA | 87 |
| 40 | LAMIACEAE | 260 |
| 40 | LAMIUM | 262 |
| 70 | LAPORTEA | 464 |
| 17 | LAPPULA | 117 |
| 13 | LAPSANA | 87 |
| 42 | LARDIZABALACEAE | 269 |
| 4 | LARIX | 19 |
| 34 | LATHYRUS | 220 |
| 42 | LAURACEAE | 269 |
| 19 | LEAVENWORTHIA | 128 |
| 23 | LECHEA | 162 |
| 54 | LEERSIA | 344 |
| 42 | LEMNA | 270 |
| 42 | LEMNACEAE | 270 |
| 42 | LENTIBULARIACEAE | 271 |
| 40 | LEONURUS | 262 |

| | | |
|---|---|---:|
| 19 | LEPIDIUM | 128 |
| 54 | LEPTOCHLOA | 344 |
| 19 | LESPEDEZA | 221 |
| 34 | LESPEDEZA | 221 |
| 19 | LESQUERELLA | 129 |
| 13 | LEUCANTHEMUM | 88 |
| 67 | LEUCOSPORA | 444 |
| 13 | LIATRIS | 88 |
| 7 | LIGUSTICUM | 34 |
| 47 | LIGUSTRUM | 301 |
| 43 | LILIACEAE | 271 |
| 43 | LILIUM | 276 |
| 44 | LIMNANTHACEAE | 280 |
| 44 | LINACEAE | 280 |
| 67 | LINARIA | 444 |
| 42 | LINDERA | 270 |
| 67 | LINDERNIA | 445 |
| 20 | LINNAEA | 147 |
| 44 | LINUM | 281 |
| 48 | LIPARIS | 313 |
| 37 | LIQUIDAMBAR | 249 |
| 45 | LIRIODENDRON | 285 |
| 17 | LITHOSPERMUM | 117 |
| 20 | LOBELIA | 144 |
| 44 | LOGANIACEAE | 283 |
| 54 | LOLIUM | 344 |
| 20 | LONICERA | 147 |
| 44 | LORANTHACEAE | 284 |
| 35 | LOTUS | 222 |
| 47 | LUDWIGIA | 307 |
| 35 | LUPINUS | 223 |
| 39 | LUZULA | 259 |
| 21 | LYCHNIS | 151 |
| 68 | LYCIUM | 457 |
| 68 | LYCOPERSICON | 457 |
| 3 | LYCOPODIACEAE | 12 |
| 3 | LYCOPODIUM | 13 |
| 40 | LYCOPUS | 263 |
| 59 | LYSIMACHIA | 380 |
| 44 | LYTHRACEAE | 285 |
| 44 | LYTHRUM | 285 |
| 46 | MACLURA | 294 |
| 13 | MADIA | 92 |
| 45 | MAGNOLIA | 285 |
| 45 | MAGNOLIACEAE | 285 |
| 43 | MAIANTHEMUM | 277 |
| 48 | MALAXIS | 313 |
| 62 | MALUS | 400 |
| 45 | MALVA | 289 |
| 45 | MALVACEAE | 286 |
| 5 | MANFREDA | 25 |
| 40 | MARRUBIUM | 263 |
| 45 | MARTYNIACEAE | 291 |
| 9 | MATELEA | 46 |

| | | | | | | |
|---|---|---|---|---|---|---|
| 14 | MATRICARIA | 92 | | 47 | OENOTHERA | 308 |
| 2 | MATTEUCCIA | 6 | | 47 | OLACACEAE | 300 |
| 43 | MEDEOLA | 277 | | 47 | ONAGRACEAE | 302 |
| 35 | MEDICAGO | 230 | | 2 | ONOCLEA | 6 |
| 14 | MEGALODONTA | 93 | | 14 | ONOPORDUM | 94 |
| 67 | MELAMPYRUM | 445 | | 17 | ONOSMODIUM | 118 |
| 43 | MELANTHIUM | 277 | | 3 | OPHIOGLOSSACEAE | 15 |
| 45 | MELASTOMATACEAE | 292 | | 3 | OPHIOGLOSSUM | 15 |
| 54 | MELICA | 344 | | 19 | OPUNTIA | 139 |
| 35 | MELILOTUS | 230 | | 48 | ORCHIDACEAE | 310 |
| 40 | MELISSA | 263 | | 43 | ORNITHOGALUM | 277 |
| 25 | MELOTHRIA | 173 | | 49 | OROBANCHACEAE | 316 |
| 45 | MENISPERMACEAE | 293 | | 49 | OROBANCHE | 317 |
| 45 | MENISPERMUM | 293 | | 32 | ORTHILIA | 194 |
| 40 | MENTHA | 263 | | 54 | ORYZOPSIS | 346 |
| 45 | MENYANTHACEAE | 293 | | 7 | OSMORHIZA | 36 |
| 45 | MENYANTHES | 293 | | 3 | OSMUNDA | 15 |
| 17 | MERTENSIA | 117 | | 3 | OSMUNDACEAE | 15 |
| 14 | MIKANIA | 93 | | 16 | OSTRYA | 112 |
| 54 | MILIUM | 345 | | 49 | OXALIDACEAE | 317 |
| 67 | MIMULUS | 445 | | 49 | OXALIS | 317 |
| 21 | MINUARTIA | 151 | | 32 | OXYDENDRUM | 194 |
| 46 | MIRABILIS | 298 | | 7 | OXYPOLIS | 36 |
| 64 | MITCHELLA | 417 | | 8 | PANAX | 42 |
| 66 | MITELLA | 433 | | 54 | PANICUM | 346 |
| 21 | MOEHRINGIA | 152 | | 49 | PAPAVERACEAE | 318 |
| 5 | MOLLUGO | 26 | | 70 | PARIETARIA | 464 |
| 40 | MONARDA | 263 | | 66 | PARNASSIA | 434 |
| 32 | MONOTROPA | 194 | | 21 | PARONYCHIA | 152 |
| 46 | MORACEAE | 293 | | 14 | PARTHENIUM | 94 |
| 46 | MORUS | 294 | | 72 | PARTHENOCISSUS | 471 |
| 54 | MUHLENBERGIA | 345 | | 54 | PASPALUM | 347 |
| 43 | MUSCARI | 277 | | 50 | PASSIFLORA | 321 |
| 17 | MYOSOTIS | 118 | | 50 | PASSIFLORACEAE | 321 |
| 60 | MYOSURUS | 385 | | 7 | PASTINACA | 36 |
| 46 | MYRICACEAE | 294 | | 17 | PAULOWNIA | 113 |
| 37 | MYRIOPHYLLUM | 249 | | 67 | PEDICULARIS | 449 |
| 46 | NAJADACEAE | 297 | | 1 | PELLAEA | 2 |
| 46 | NAJAS | 297 | | 8 | PELTANDRA | 41 |
| 45 | NAPAEA | 289 | | 67 | PENSTEMON | 450 |
| 19 | NASTURTIUM | 130 | | 66 | PENTHORUM | 434 |
| 46 | NELUMBO | 299 | | 7 | PERIDERIDIA | 36 |
| 7 | NEMOPANTHUS | 40 | | 41 | PERILLA | 264 |
| 41 | NEPETA | 264 | | 68 | PETUNIA | 457 |
| 19 | NESLIA | 130 | | 38 | PHACELIA | 251 |
| 68 | NICANDRA | 457 | | 55 | PHALARIS | 348 |
| 43 | NOTHOSCORDUM | 277 | | 35 | PHASEOLUS | 232 |
| 46 | NUPHAR | 299 | | 66 | PHILADELPHUS | 434 |
| 46 | NYCTAGINACEAE | 297 | | 55 | PHLEUM | 348 |
| 46 | NYMPHAEA | 300 | | 56 | PHLOX | 360 |
| 46 | NYMPHAEACEAE | 299 | | 44 | PHORADENDRON | 284 |
| 46 | NYSSA | 300 | | 55 | PHRAGMITES | 348 |
| 46 | NYSSACEAE | 300 | | 70 | PHRYMA | 467 |
| 37 | OBOLARIA | 245 | | 70 | PHYLA | 467 |

133

135